THE BEGETTERS AND THE BEGOTTEN

WHENCE AND HOW CAME WE ALL?

by

Thomas A. Leonard, M.D.

edited by

Marjorie Heiner Bartlett

Medical Physics Publishing Corporation
Madison, Wisconsin

1988

Published by:

Medical Physics Publishing Corporation
B428, 1300 University Avenue
Madison, Wisconsin 53706

ISBN: 0-944838-01-4

Library of Congress Cataloging in Publication Data:

Leonard, Thomas A., 1896-
The begetters and the begotten: whence and how came we all? / by Thomas A. Leonard; edited by Marjorie Heiner Bartlett.
p. cm.
Bibliography: p.
ISBN 0-944838-01-4 (pbk.) : $20.00 (est.)
Life--Origin. 2. Evolution. 3. Human evolution. 4. Life (Biology) I. Bartlett, Marjorie Heiner. II. Title.
QH325.L45 1988
577--dc19 87-34948
CIP

Note: The chapters of this book have been reviewed and updated by authorities in each subject area, and all possible care has been taken to ensure the accuracy of the material presented. Given the extremely broad scope of the material covered and the inability to prove or disprove some of the author's viewpoints, the editor and publisher assume no responsiblity for the contents.

Cover Design Concept: Marjorie H. Bartlett

Cover Artist: Michael G. Smith

DEDICATION

I wish to dedicate these writings to the memory of my mother, the late Caroline Leonard.

During a lifetime of 99 years, she gave a continuing demonstration of the precious qualities of selflessness, kindness, and love beyond measure. For her six sons and two daughters she set a wonderful example of tolerance and understanding of the beliefs and actions of other people, even when they differed widely from her own. And to each of us she imparted a deep appreciation of work and learning, not only as essentials to self-reliance; but, as a means of contributing to the happiness of others.

No tribute or honor I might bestow could adquately express my love and admiration for her.

Thomas A. Leonard, M.D.

Caroline Leonard
September 19, 1868-December 5, 1967

FOREWORD

Awaiting the reader of this book is a fascinating panoramic account of life from the beginning to the present. The most fundamental and the most recent discoveries are presented in language and imagery that a layman can grasp and enjoy.

The author's outlook is decidedly scientific, but his appreciation of nature is reverential. It springs from a deep, quiet religious faith. In addition to his unique perspectives, he has an amazing ability to guide the reader expertly through the awesome mysteries of creation and evolution.

Dr. Thomas A. Leonard is a retired physician, a nonogenarian, with decades of devotion as a gynecologist and obstetrician. Throughout his professional career he has been known as a caring, compassionate physician who tenderly delivered thousands of babies. He was also a diligent researcher engaged in continuing study and exploration, always learning and sharing. His studies of the health problems and needs of the expectant mother produced a dramatic decline in maternal mortality and a consequent rise of infant survival. For these and countless other contributions to the enrichment and enjoyment of life, he was accorded high honors and widespread recognition, yet his most obvious and endearing characteristic is a deep sense of humility.

During the past several years, Dr. Leonard has been virtually sightless--he sees only shadows. Yet he "reads" eagerly and enthusiastically, seizing full advantage of the great wealth of information available on sound recordings. This book was spoken into a tape recorder and then edited. One can only conclude that the author's loss of sight sharpened his vision and enabled him to grasp, encompass, and transmit to us his wonderful view of life in all of its majesty.

Harold B. McCarty (deceased), Late Emeritus Professor and
Director, Wisconsin Public Broadcasting Service,
University of Wisconsin-Madison

"Standing Mother and Child" ceramic, 1976 from the collection of Professor and Mrs. Michael Petrovich. One of a series of mother and child statues by Dr. Thomas A. Leonard.

ACKNOWLEDGEMENTS

I am deeply indebted to many individuals and institutions for helping make this book possible.

For a great deal of highly technical and historical information, I owe much to several faculty members of the University of Wisconsin-Madison: Arthur B. Chapman, Emeritus Professor of Genetics; Lowell R. Laudon, Emeritus Professor of Geology; Dr. John D. Ferry and Dr. John Willard, Emeritus Professors of Chemistry; Menahem Mansoor, Emeritus Professor of Hebrew and Semitic Studies; and the late Herbert Clarke, Professor of Botany. With kindness and generosity they patiently answered my questions and supplied valuable source materials and references.

The National Library Service for the Blind, through its Regional Library in Milwaukee, was particularly helpful. They promptly provided either disc or audiocassette recordings of the many books I requested and sent, in addition, much related material. Recording for the Blind, a nonprofit agency in New York City with volunteer tapists throughout the country, supplied similar services, all with personalized attention and speedy delivery.

Numerous individuals read source material to me. I wish to thank, in particular, Mrs. Jean Soper of Sanibel Island, Florida; Mr. Harry Sellery of Middleton, Wisconsin; Mr. Dean Ball of Westport, Wisconsin; and Miss Elena Meyer of Madison, Wisconsin.

The State Medical Society of Wisconsin, through its executive secretary, Mr. Earl Thayer, and assistant secretary, Mr. William Wendle, gave important help in the initial stages of converting my sometimes rambling dictation to orderly typed copy. Ralph Hawley, Associate Dean of the University of Wisconsin School of Medicine and Executive Director of the U.W. Medical Alumni Association, has been of great assistance in many ways. Transcribers Terri Weaver and Vicki Meyer deserve special mention. I thank Mrs. Gundega Korsts for help in the original manuscript preparation.

For my wife, Myrtle, my thanks are boundless. She read to me often, consulted the dictionary and encyclopedia endlessly, transported me to many places to obtain source materials, and listened painstakingly to my original dictations and many refinements. She offered much constructive criticism and provided needed relaxation, along with tea and snacks, during the long hours I was confined to my study for listening and recording. She has been truly a loving helpmate.

Special, heartfelt thanks go to my good friends, Dr. Michael B. Petrovich, Professor of Russian and Balkan History at the University of Wisconsin-Madison, and the late Professor Harold B. McCarty, formerly Director of the public radio and television services of the University and the State of Wisconsin. Michael gave guidance, encouragement, and inspiration throughout the entire project. "Mac" gave me needed counsel on both oral and written communication.

I am also appreciative of the assistance of Marjorie Bartlett, who brought many years of experience to the task of a final editing of these chapters. Marje succeeded in retaining the flavor of my writing as she strove to make words, phrases, and clauses contribute to the intended impact of the manuscript. It is my opinion that without the help of Marje this book would never have been published. Dr. John Cameron, Emeritus Professor of Medical Physics of the University of Wisconsin-Madison and founder of the Medical Physics Publishing Corporation has provided the essential impetus and means for the final publishing of this work. Pam Haney chose and arranged for many of the pictures appearing in the text. Many people reviewed the text for accuracy of technical data; among them I would like to thank Dr. James Crow, Dr. Arthur Hasler, Dr. Mary Jo Donovan, Mark Slovak, Dan Peterson, Bettina Shell, and Bette Barnes.

There are many others who in some way, large or small, consciously or unconsciously, have aided in the completion of this work, and I am deeply grateful to them all. I have learned that no one ever works alone. Others are always involved and deserve a large share of the credit for any accomplishment.

Thomas A. Leonard, M.D., F.A.C.O.G.

November 1987

A NOTE FROM THE EDITOR

Perhaps it is unusual for the editor of a book to write a message to the readers. But, this is an unusual book. This is Dr. Tom Leonard's book.

I feel very privileged to have been asked by Dr. Leonard and the publisher, Dr. John Cameron, to undertake the editing of the manuscript. I want the reader to feel the nearness, the closeness of Dr. Leonard. Picture him sitting by a large window looking out over Wisconsin's lovely Lake Mendota -- a view he must see, now, in his mind's eye since he has been without sight these last ten years.

In editing this manuscript, I have tried to retain, throughout, the special love and awe that Dr. Leonard transmits. His personal friends will, I hope, say to themselves, "Now, doesn't that sound just like Tom Leonard!"

As Dr. Leonard talks, I visualize my listening intently and, every so often, interjecting a hushed, "Amazing, simply amazing!", as his story unfolds. And this is the reaction Dr. Leonard hopes will be that of the reader.

Not a textbook, this book is one to be read for pure enjoyment and for expanding one's thoughts. The material presented forms a fine framework upon which to attach newly acquired facts, contemporary science and medical reports, mathematical and philosophical discourses, new theories and educational experiences. In short, after reading Dr. Leonard's book, things fit together. During one's schoolyears and later life, knowledge is acquired in separate subject clumps. Now, they blend and one can appreciate the wholeness of all that exists.

As I write, our ship among others from many continents, is anchored for several days off the city of Paranagua, Parana, Brazil. A 360° vista of sea, green pudding shaped islands, lush lowlands, sky with great gray low scudding clouds, beams of sun, rain and rainbows, ships and more ships from all over the Earth, straining and arcing at anchor, frigate bird, porpoise and human – the interrelationship becomes apparent. And, it is the undeniable uniqueness and value of humans that Dr. Leonard emphasizes – the uniqueness of each of us and of you, the reader.

Marjorie Heiner Bartlett

About the Editor...

Marjorie Heiner Bartlett was, for seventeen years, Administrator of HEALTH-LINE, a program of the Department of Health and Human Issues, University of Wisconsin-Madison and U.W. Extension.

Her undergraduate studies were in Zoology at the American University in Washington, D.C. and the University of Wisconsin-Madison. She holds a Master of Philosophy degree in Zoology-Cytology. Her biological and medical illustrations have been published in journals and used in classroom presentations.

Marge has travelled widely including a number of adventurous trips on cargo vessels. She has seven published articles.

PROLOGUE

> "We shall not cease from our explorations, and the end of our exploring will be to arrive at where we started and to know the place for the first time."
>
> T.S. Eliot
> Little Gidding, from Four Quarters

This book is about life.

It is concerned with that mysterious quality which causes atoms and molecules when properly assembled, to function in a most unique manner. Stirred by the spark of life, inanimate matter becomes animate. The nonbiological world becomes biological. With the advent of life, planet Earth could no longer remain the sterile globe it once was. It became abundantly endowed with an infinite variety of structures and forms. The plant and animal kingdoms came into existence.

It is the latter, the animal kingdom, with which these writings will mainly deal. Let us meet a group of mythical travelers and accompany them along the pathway of the immense journey (as it was so aptly named by the great naturalist, the late Loren Eiseley). We will follow our travelers' progression through four phases: The Beginning of Planet Earth, The Advent of Life on Planet Earth, The Proliferation of Life on Planet Earth, and Humans Come to Planet Earth.

Part I, The Beginning of Planet Earth, has to do with the creation of the universe, the formation of suns and galaxies, and our solar system. Part II, The Advent of Life on Planet Earth, looks at chemistry and physics, at the electromagnetic energies and their interrelationships and the roles they may have played in the creation of living matter. Part III, Life Proliferates on Planet Earth, deals with the many great evolutionary advances, from the simplest to the most complex of living organisms. Part IV, Humans Come to Planet Earth, traces the evolutionary concept through the primates to the emergence of human beings.

This book is not presented as a profound, specialized, academic work. It is intended to be enjoyed and absorbed by any interested reader. Its writing has been motivated by the nature of the author's profession. As a specialist in gynecology and obstetrics, he has frequently been called upon to assume the almost sacred responsibility of providing the maximum safety and security for both mother and child during the birthing process. To hold a little newborn, recently deprived of its near-marine existence, to observe that first intake of

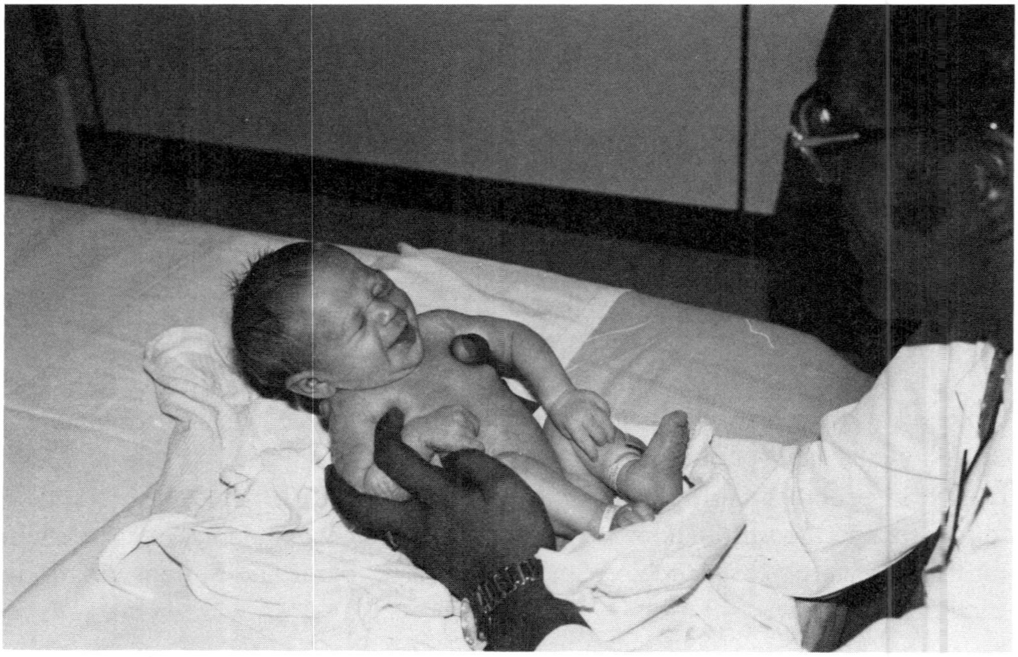

Baby Hannah Manthey, five days old, in the hands of Perry Henderson, M.D., Professor of Obstetrics and Gynecology, University of Wisconsin.

breath, to watch the struggle, to imagine what it has seen with the first opening of its eyes -- these and many more first reactions have always had a strong and deep effect upon the thoughts and the reflections of the author. Each such incident was not only what it seemed to represent at the moment of birth, but a reminder that the infant's evolution extended far back into cosmic time.

Some of the answers to the problems of life may seem far beyond our ever knowing. But let us deal with that which has become known throughout the ages, and especially during the last few decades. It is truly a fascinating story. Much factual information is used. However, the data are not complete and where there are occasional gaps the author and readers must resort to conjecture. Interspersed throughout the book, readers will find personal observations and philosophies of the author, often reflecting his professional life.

The dominant motivation has been to construct a philosophy of the origins and the perpetuation of life which may be consistent from both metaphysical and scientific points of view.

Occasionally, the author has intentionally used repetition in order to emphasize a point or to stress continuity. If there has been some lapse, some omission, some falling short of perfection, the author begs forgiveness, understanding, and tolerance. Being human, he does not have that flawlessness which the gods alone are supposed to possess.

Respectfully, Thomas A. Leonard, M.D., F.A.C.O.G.

TABLE OF CONTENTS

Part 1: FROM COSMOS TO PLANET EARTH

Chapter 1: THE ORIGINS

It is a beautiful spring day. The patio is flooded with brilliant sunlight. Flowering plants are beginning to emerge and some are even now displaying their multitudinous colors and forms. What a time for creation! The sculptress sits before her bushel-sized piece of stone. It is irregular in shape, coarse of texture, and dirty grey in color. Previously, she hoisted it into place by means of a chain sling hooked to a rope pulley supported from a derrick. The stone is placed at a convenient height for working. The chisels and the points are sharpened and laid into place along with the hammers and mallets. Despite the readiness of the tools and of the stone, she is not yet fully prepared to begin her work. She will examine this stone from all angles and peer into its utmost depths. This may go on for hours, days, weeks, or even months before she discovers that which could lie within. Only then, by her special arts and skills, will she be able to bring forth that refinement, that beauty of color, texture, and bond which she has envisioned.

Michaelangelo, "Bearded Captive." One of a series of unfinished sculptures. Marble, Florence Academy, Italy. 1527-28.

So it is with the potter working with formless, moist clay, drawing out of the meaningless mass things of great beauty. It is true, also, of the sculptor working in ceramics who can elicit from formless plastic material a portrayal of the profound love which exists between mother and child. The potential for beauty and form, emotion and message have always been present in base materials. Their release depends upon the artist – Michelangelo, da Vinci, Mestrovic, Carl Milles, Rodin, Henry Moore, Picasso, or one of numerous others who has been endowed with the ability to bring out that which is perceived to be within.

Isn't all of creation something like this? Hasn't the potential for everything always existed from the beginning? Have we lacked the ability to recognize this potential because its substance was in a different form from that which now characterizes its existence?

One of the most profound of all natural laws is that matter can be neither created nor destroyed. Matter can be changed. If, in the process of change, any loss appears to have occurred, this can be fully taken into account by the production of energy. The total amount of matter contained within the universe is constant – has always been, is now, and ever shall be.

The universe consists of units of matter, and matter can change itself into other forms of matter or energy. The atom is the building unit of all matter, animate and inanimate. Each atom contains a specific number of protons, neutrons, and electrons. Each atom carries a positive electric charge in its nucleus and usually the same amount of negative electric charge in its surrounding electrons. Atoms join with other atoms to form molecules.

Formation of water is brought about by the union of two hydrogen atoms with one oxygen atom. By employing certain highly technical procedures, the nucleus of the atom can be split, producing tremendous heat, light and radioactivity, such as is manifest by the explosion of an atomic bomb. When the nuclei of light atoms fuse, they produce heat. When the nuclei of two hydrogen nuclei (protons) unite, the resulting matter is not double that of one proton; but, the combined matter is reduced slightly. If we apply the natural law that matter can neither be created nor destroyed, the result of such a fusion must include the energy produced – heat, light, and electromagnetic energy.

In the beginning, the potential for all of the processes which have culminated in the universe as we experience it – the billions upon billions of stars, ours and numerous other solar systems, with our Earth and all that it contains – resided in a mass of hydrogen atoms. If we wish to go back to the roots of our origin, we must find a means of retreating into this remote past, to the instant of the beginning. Just as artists can penetrate beyond the surface areas of stone or clay to bring forth what is hidden, so it is with our researches into our origins. The past can be likened to stone or clay. Surface observation does not and cannot reveal all that is concealed within. If we are to reach into the past, we must have sophisticated instruments and technologies and we must have scientists with the capabilities to use such instruments and technologies and to interpret and apply the collected information.

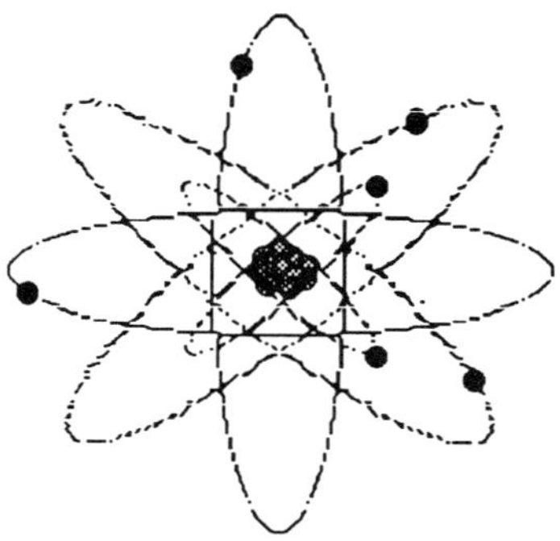

Diagram of an atom, showing the nucleus, made up of protons and neutrons. The nucleus has a positive charge, and attracts the negatively charged electrons, thus holding the atom together. The nucleus contains almost all of the atom's mass, but its size relative to the electrons' orbits is much smaller than shown here.

Fortunately, we do have both the technologies and the minds. Research has taken us far. Radiometric techniques have enabled scientists to date the age of the Earth at 4.6 billion years. By means of the radiotelescope, observations can be made far beyond our own galaxy. And, by means of huge, dish shaped receptors, radio signals from interstellar space are being captured. Interpretation of these signals has led to measurements of cosmic time. The age of the universe is now set at 18 billion years.

Most of this knowledge has come to us only within the last three decades. It is not difficult to assume that what lies ahead will be even more revealing. This is a far cry from what must have been the understanding of earliest men, whose knowledge was severely limited to what the senses could reveal.

The early humans could see only as far as the horizon, and could touch only that which could be reached by an extended arm or with a simple tool. They could feel cold and warmth, pain and pleasure, and hunger. They must have been awed by flashes of lightning and the roar of thunder, and by the accompanying deluge of rain. They must have viewed with wonder the stars of the sky, the periodic appearances of the moon, the rising and setting of the sun and its travels from one horizon to the other. All of these must have been viewed with a variety of emotional responses – sometimes happiness, sometimes fear. The ability to reason and to assign cause-and-effect relationships developed over a long time, beginning we know not when, and continuing to develop into, and, we hope, beyond the present.

Indeed, mankind has come a long, long way since the cave dwellers. Should it be assumed that the end of knowledge is here or just around the corner, however, let us think again. It is true that much has been learned, and much accumulated knowledge can be interrelated and interpreted, but what has been accomplished so far is only a beginning. It is minimal compared to what would be required for mankind to truly understand his origin, his evolution, his being, and his destiny.

Now we want to learn, insofar as possible, how we can relate ourselves to all that is and ever was. The fascinating story unfolds.

Part 1: FROM COSMOS TO PLANET EARTH

Chapter 2: THE UNIVERSE

The universe is a phenomenon of space, time, substance and of gravitational and electromagnetic forces. Within its boundlessness, so incomprehensibly vast, there exists all matter and energy that ever was, is now, and ever shall be. There is nothing other than this. As to where, when and how it all began, people have many different ideas. Some deny that there ever was a beginning. To most of us, however, the universe remains marvelous, miraculous, and mysterious.

The components of the universe are so numerous that it has been impossible to count, measure, study, and classify all of them. The great advances of recent years of space exploration have helped us in our study. Scientists have been persistent in their search for knowledge and understanding. By their ingenuity and imaginative powers, they have developed instruments and technologies which permit penetration into the utmost regions of space, to distances which can be measured only in terms of billions of light years. (A light year is the distance that light, traveling at the speed of 186,000 miles per second, covers in one year. This enormous distance, multiplied to hundreds, thousands, millions, and billions, makes the size of the universe seem indeed overwhelming.) But even with all of these advances, a total concept of the universe is not within our grasp. The unravelings have just begun.

In presenting theories and conjectures, credibility must be established on the basis of all facets of biological and nonbiological evolution, and we must abide by universal natural laws. Since we as human beings, with our atoms, molecules, and compounds, have evolved from and are now a part of everything that ever was, the story of our evolution too must start at the very beginning.

There is controversy over whether the universe had a beginning, or whether it has always existed. Some believe that after long periods of expansion and contraction it repeatedly dies and is reborn, forming a new universe. No one knows. Assuming, however, that there was a beginning, at least for the cycle that the universe is now in, we can examine one of the current theories about how the universe might have been formed.

We marvel at ourselves as products of cosmic time and of all that the universe contains. We view with wonderment and amazement the constant changing of each component of the universe, considering that each positional change must alter gravitational and other courses, for one component relates to every other. Such exactitudes, and there are many, will be dealt with many times as this story advances. Now, we ask, "Can all of this precision be accounted for on the basis of mere happenstance? Is the universe a plaything of the gods, created in order that mortal man may be confused? Or, is it part of a divine plan which has been designed and coordinated, revealing little and concealing much?" Questions such as these must be deferred until the end of this book. Even then, they must be answered by each reader individually.

The universe began with an explosion – not the kind of localized explosion we think of normally, but a huge fireball that filled all of the universe. This is what scientists call the Big Bang. Present were electrons – negatively charged subatomic particles, and positrons, which are positively charged particles with precisely the same mass as an electron. Combinations of electrons and positrons are known as matter and antimatter respectively. Also present were neutrinos and light. Light is made up of photons, which are formed by the annihilation of matter and antimatter. This process can also be reversed – photons create matter and antimatter. Both processes can only take place under conditions of intense heat and light, such as were present in the first tiny fraction of a second of the Big Bang.

As the universe aged, it cooled. When it was less than a second old, heavier particles such as protons and neutrons were created. After about 100 seconds, particles began to combine to form hydrogen nuclei. During the next few minutes, about one fourth of the mass of the universe formed into helium, while almost three fourths of the mass consisted of hydrogen.

For the next few hundred thousand years, the fireball was much like the center of a star – hot and opaque, with radiation bouncing from particle to particle.

One half million years after the Big Bang, the universe became transparent. Free electrons were drawn into nuclei, and the first atoms were formed. Matter no longer stopped radiation, including light. The radiation could pass through the matter. Matter and radiation no longer interacted and were then separate.

One billion years after the Big Bang, hydrogen and helium began to condense under the influence of gravity. Stars and galaxies began to form. As stars condensed, they became hot and began to shine. Nuclear reactions were ignited, and the gradual synthesis of the heavier elements began.

The evolution of galaxies can be compared to the generations of humans – older galaxies continually die out, and new ones are being formed. Our sun and its planets, including Earth, were formed coevally (at the same time) about 4.6 billion years ago. This time is based on radiometric measurements of the age of the Earth.

Everything that exists today is a consequence of the Big Bang. As Professor George Abell of UCLA wrote in his book *Realm of the Universe:* "The fireball is still existing, in a sense. It has expanded greatly, but the original matter and radiation are still present and accounted for. The stuff of our bodies came from material in the fireball. We were and still are in the midst of it; it is all around us."

Many fascinating aspects of the universe have been revealed in the course of research into the Big Bang theory. For instance, it has been concluded that the substance of the newly formed sun and of the Big Bang were identical.

Dr. Melvin Gottlieb, Research Professor of Plasma Physics at Princeton University, has conducted some unique experiments in which he has been able to fuse hydrogen nuclei and in the process create, for a fraction of a second, the almost unbelievable temperature of over 60 million degrees celsius (°C). His process was that of depriving hydrogen atoms of their electrons, thereby ionizing them. Then the hydrogen ions were suspended in a vaporized plasma of gaseous ions and passed through a magnetic field. There the fusion occurred. In the August 28, 1978 issue of *Newsweek* magazine, Dr. Gottlieb stated that he believed that, with refinements of technique, a temperature of over 100 million °C could be obtained. It is interesting to note that at these extremes of temperature the energetic radiation produced is invisible.

Another interesting aspect of the universe is that, since the time of the Big Bang, the various components have been constantly receding from their original source. Robert Jastrow relates in his book *God and the Astronomers*, how the American astronomer Edwin Hubble, in 1929, established that the more distant a celestial body, the more rapidly it recedes from Earth – an empirical formulation now known as the Hubble-Humason Law.

When American astronauts departed from the moon, they left behind a number of scientific instruments so that remote observations could continue. One of these instruments was a reflecting mirror which proved highly useful to Harlan Smith of the University of Texas. Using laser beams directed at the mirror (from the 107 inch telescope at McDonald Observatory) and reflected back to Earth, Smith has measured accurately, over the course of several years, the distance from a given point on Earth to a given point on the moon. He determined that the moon is receding from Earth at the rate of 1 centimeter per year.

8 Although this is due to tidal friction (another type of gravitational interaction between the moon and the Earth), and not to expansion from the Big Bang, this experiment is a good example of how such questions are approached.

Still further credence is given to the Big Bang theory by some fascinating results obtained by two 1978 Nobel Prize winners in physics. Dr. Arnold Penzias and Dr. Robert W. Wilson of the Bell Research Laboratories in New Jersey have been long-time champions of the theory that the universe began with a gi-gantic explosion. They constructed at Homebel, New Jersey, special devices designed to receive microwave radiation. They discovered microwave radiation arriving from space, and they were able to show that this radiation was a fossil relic from the early universe. In fact, this radiation represents cosmic time, since it is thought to result from the radiation created at the time of the Big Bang. This beginning, according to Penzias and Wilson, and confirmed by other investigators in this particular realm of research, is now estimated to have been 15-20 billion years ago.

The other planets of our solar system originated at the same time as the Earth. Furthermore, samples brought back from the moon by the astronauts and the sample probings of Mars by the Viking spacecraft show that at least these two extraterrestrial bodies are composed of the same materials as Earth. There are impact craters on the moons of Mars – Phobos and Deimos – indicating that they, too, received meteorites from space. Analyses of the atmospheric gases on the moon compared to gases on the Earth indicate similarities there as well, for both contain carbon dioxide, hydrogen sulfide, and nitrogen.

Some of the planets are nearer to the sun than Earth, and some are farther away. For example, Neptune is 2.5 billion miles from the sun, while Pluto is even farther away, at 4 billion miles distant. But despite the many similarities of the planets of our solar system, there is one great difference. Insofar as has been determined, none other than Earth currently has the capability to support life.

If the same process which guided the evolutionary directions of our sun in producing its solar system applies throughout the universe, then many of the stars, in all probability, have done likewise. Thus, we have in our universe a wonderful magnificence of substance, of energy, motion, heat, light, and radioactivity, of atoms and molecules – all taking part in a tremendously complex interplay which has resulted in the potential for what many believe to be the ultimate goal of creation, human life. The following chapters will attempt to examine in detail the various components of the universe and their individual and related roles in the evolutionary development of the ultimate purpose of creation.

It all began with the fusion of hydrogen nuclei.

Part 1: FROM COSMOS TO PLANET EARTH

Chapter 3: THE SUN

When one looks at a fine tapestry from the reverse side, all one sees is a jumble of unclipped or knotted threads, a blur of colors, but nothing of what lies on the proper side of the fabric – the magnificence, the beauty of workmanship, the message, and the creative skill of the artist.

In tapestry, differences in perception lie within a few millimeters of each other. Even if one could see only the reverse side of the tapestry, by tracing the individual threads, and after much tireless effort interrelating them, it may be possible finally to form a concept of the right side.

The universe is, perhaps, such a tapestry of chemical, physical and electromagnetic forces. It, too, has a magnificence, a creative beauty and almost certainly a purposeful message, but a great portion of this is obscure and hidden from us. Scientists must find and isolate each individual thread, follow it through, and make the interrelationships necessary for understanding. Such penetrations into this fabric of time and space are the only means by which we can come to understand the universe.

A glance into the sky on a cloudless, moonless night reveals myriads of twinkling lights, each of which is a star not unlike the center of our solar system. And those we can see are only a few of the hundreds of billions that exist. These stars formed from great clouds of hydrogen gas within the galaxy. Each star is made up of a vast hot sphere of hydrogen gas. At the center the hydrogen burns, and is thereby transmuted to heavier elements. Stars group themselves into vast clusters called galaxies, each of which contain hundreds of billions of stars revolving around a common center. The universe as a whole contains billions of galaxies. Each is in the process of moving away from its neighbors as the universe expands. This expansion is a result of the explosion which began our cosmos 15 billion years ago.

Most suns have a burning life of billions of years. Many suns have already used up all the available fuel in their centers and, in the process, have collapsed into slowly cooling ancient white dwarf stars. Sometimes a star will end its life with a massive explosion called a supernova, whose brilliance can even outshine a galaxy. The remains of the star quickly collapse into a rapidly rotating, incredibly dense sphere with a diameter of about 10 kilometers. If the remnant is massive enough, even atoms cannot resist the gravitational crush. The remnant will collapse so completely that despite having a finite mass, it will have no volume, and will exist as a mathematical point. Its gravitational field will approach infinity, so strong that not even light can escape it. This phenomenon, known as a black hole, has not yet been observed, but its existence is predicted by the theories of modern physics.

The distances between stars are so immense that they are measured in terms of light years, or the distance which light travels in one year. Measurements by means of the radio telescope indicate that the Milky Way Galaxy is approximately 100,000 light years in diameter. If we view a star on the margin directly opposite from where our sun is located, we are seeing light which was generated 100,000 years ago. If we send a signal which travels at the speed of light, it will require 100,000 years to arrive at that star.

Our sun is 860,000 miles in diameter and is 93 million miles from the Earth. Spectographic examination reveals that it is a conglomerate of "burning" substances, that the heat, light, and radioactivity which are generated there result from fusion of hydrogen nuclei in the core, and that central temperatures reach millions of degrees. It is believed that our sun still has within itself adequate energy reserves to burn for another 5 billion years.

The sun has multiple effects upon Earth. The gravitational effect of the sun keeps Earth and the other planets of our solar system in their respective orbits. Since Earth makes a complete rotation on its axis each 24 hours, all parts of the Earth are exposed to the effects of the sun for an average of half of that time (except for the polar regions, where seasonal exposure to the sun varies more due to the tilting of Earth on its axis). This means that the ecological benefits of the sun are available to all parts of Earth.

Electro-magnetic spectrum, showing differences in frequencies and wavelengths of the several types of electro-magnetic waves.

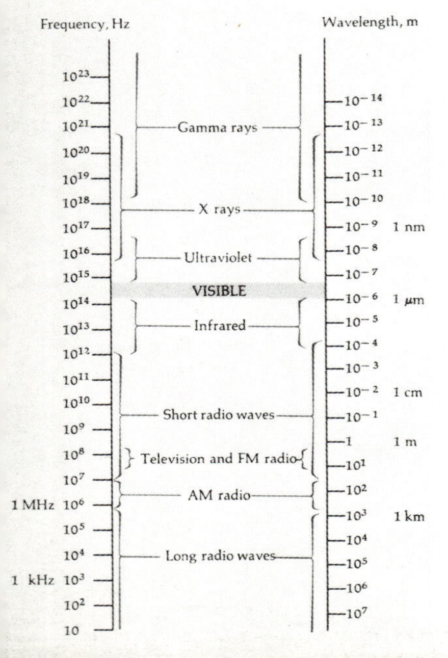

Light rays enable plants to engage in the marvelous process of photosynthesis, whereby they utilize water and carbon dioxide from the air, and convert them into oxygen and carbohydrates. This oxygen forms the basis of our atmosphere. Plants are the basis for the Earth's food chain. Without plants, all living creatures would die.

Humans could not appreciate the beauties which abound upon the surface of the Earth were it not for the light rays of the sun. The human optical system is so well adapted to the available light that we can perceive size, shape, form, color, and depth. These same rays are absorbed by the human skin and transferred to the underlying fat where vitamin D is manufactured, ensuring proper bone formation. Yet too much vitamin D and too much ultraviolet are toxic, so in races which have long resided in the tropics where the sun's rays are most intense, the skin has protective pigmentation. The effects of the sun are extremely bountiful, but they can also be destructive. Too much exposure, especially among those who are lightly pigmented, can result in severe burning of the skin and even in the formation of cancer. With the exhaustion of fossil fuels, it may be necessary for humans to depend more and more upon the sun for fuel needs.

What might have been the thoughts of the first humans when they observed the daily rising and setting of the sun and its travels through the sky? In many early mythologies, the sun was considered to be a supreme god, or the sun was believed to be controlled by one of the most powerful gods of the pantheon.

Many rulers throughout history believed the sun to be the supreme deity. In their desire for power and immortality, they claimed that they were themselves divine, and akin to the sun. About twelve centuries before the Christian era, at Thebes in Egypt, Amon-Re was established as the national god of Egypt. Amon-Re was a sun god, representing the sun's life-giving light and warmth. Egyptian rulers identified themselves as the sons of Amon-Re.

In the December, 1978 issue of *Scientific American*, Donald Redford, Professor of Archaeology at the University of Pennsylvania, in an article titled "The Re Kingdom of Akhenaten," describes in detail his excavations at Karnak and their meaning. In the 14th Century B.C., in the 18th Dynasty, Amenhotep IV succeeded to the throne as Pharaoh. He assumed the name Akhenaten as an expression of his devotion to one god, Aten, above all other gods of Egypt. Aten was represented by the sun disc – a large golden disc with rays extending in all directions and a human hand at the end of each ray. Aten was seen as the creator of the natural world. He was depicted as a benevolent father actively concerned for the daily maintenance of all his creatures, even the lowest ones, throughout the world. With the increasing elevation of the worship of Aten, Akhenaten eventually ordered that all worship of other gods cease. Akhenaten is thus customarily credited with establishing the first monotheistic religion.

The god Aten is represented by a large golden disk over the head of Tutankhamen. Wood overlaid with gold from the tomb of Tutenkhamen, Metropolitan Museum of Art, New York.

Blocks of stone for a new temple to Aten were quarried 80 miles away from where they were erected at Karnak. After these tremendous stones were in place, bas relief carvings were made of the processions led by Akhenaten, his queen Nefertiti, and their daughters all proceeding toward the sun disc to worship and perform sacrifice. About 17 years after the death of Akhenaten, Harmhab destroyed this beautiful temple.

Worship of a sun god was also a practice of the Inca civilization in Peru and of the Aztec civilization in Mexico. The Aztec image of the sun god was an anthropomorphic figure in the form of a plumed serpent, Quetzalcoatl The Aztecs believed that the sun god demanded human blood, and that if it were not fully satiated, the sun god would wreak vengeance upon his worshipers. Thus, human sacrifice had a strong place in the Aztec religion. Parchment codices record how certain youths were schooled to become victims of this human sacrifice. They were held to the sacrificial stone by force, the chest was opened with a sharp piece of obsidian, and the beating heart was torn out and offered in sacrifice to the sun god. Thousands of prisoners of war were sacrificed in this manner.

Dr. Hazel Alberson, Emeritus Professor of Comparative Literature at the University of Wisconsin-Madison, has provided interesting insights from ancient Greek mythology. In the Greek pantheon, Helios traversed the sky daily in his chariot of fire. Phaeton, Helios' son, wanted to dispel rumors questioning the legitimacy of his parentage by proving he could ride his father's flaming chariot through the sky. Phaeton succeeded in persuading his reluctant father to allow him to do this, but he lost control of the horses. The chariot swerved, coming so near to the surface of the Earth in the vicinity of Africa that it burned and scorched the land into a barren desert, devoid of plant life, and the skins of the people who resided there became black.

Until the end of World War II, the Emperor of Japan declared himself divine and a direct descendant of the sun. Also, in the Hawaiian pantheon the sun was the supreme god of all the gods, goddesses, and demigods.

Early man could view the sun only in terms of wonderment and awe. Modern man, because of the extensive research of countless scientists and resulting interpretations, can view the sun in the light of knowledge. We now know that the sun, the center of our solar system, transmits to us on its various wavelengths of heat and light (and the rest of the electromagnetic spectrum) something of itself which becomes an integral part of ourselves, and without which we could not long survive. We are, therefore, a part of the sun just as the sun is a part of us. Likewise, both the sun and we are parts of all that the universe contains.

Use of Quetzalcoatl in building design. Detail of an 11th century wall frieze, Chichen Itza, Yucatan, Mexico. The feathers of Quetzalcoatl can be seen over and around the doorway.

Part 1: FROM COSMOS TO PLANET EARTH

Chapter 4: PLANET EARTH

The primitive person's limited view of his or her habitat did not permit knowing it in totality. Horizons were limited to what the optical senses could survey. Humans struggled with questions concerning the origin of this habitat, its shape and size, its status in the universe, its many changing aspects and the role which it was destined to play in the creation of life, its evolution and perpetuation.

The idea that planet Earth was flat, as observed by an individual standing on only a small portion of it, or the idea that Earth was the center of the universe could not be dismissed until magnified observation and mathematical deduction proved otherwise.

Ptolemy's views on astronomy, which he drew mainly from the theories of Hipparchus, remained dominant until the 16th century. Ptolemy believed that all celestial bodies revolved in circular orbits. Some planets simply revolved around the Earth; others rotated around points which in turn revolved around the Earth. The stars were points of light fixed on a dome which revolved around the Earth, attracted to it by gravity.

In 1513, Copernicus published a book in which he wrote that the sun is the center of our solar system, and that Earth and the other planets revolved around it. He died soon after the book was published.

Almost a hundred years later Galileo turned the newly invented telescope to the stars, and soon made a series of important discoveries. He found evidence that proved that the Earth revolved around the sun, and was not the center of the universe. For holding this position, he was called before the Inquisition. Listening to the agonizing cries of another man being tortured on the rack, Galileo was forced to recant, and to deny the ideas he had truthfully and conscientiously proclaimed.

The concept of a flat earth persisted and was perpetuated. It was thought possible, if one wandered too far, to fall off the edge of Earth and perish. Evil beings lived in the Antipodes, the name for the other side of Earth.

When northern barbarian tribes descended upon lower Europe, therefore, it was thought that they were evil Antipodians who crawled up over the edge of the Earth to engage in war, rape, and destruction. It is strange indeed that the minds of people can be influenced, on the one hand, by truth and enlightenment and, on the other hand, by ignorance and evil intent.

There is now proof that the Earth is a magnificent globe suspended in space – not by wires and trappings such as those employed in the construction of a mobile, but held by the invisible forces of gravity. While on their lunar journey the astronauts observed this superb phenomenon. They have not only seen this spectacle for themselves, but they have shared it on television with all who wished to observe. There can now be no denial that Copernicus and Galileo were right. Our solar system, with the sun at its center and the Earth and its sister planets revolving about it in a complex of moving, changing gravitational relationships, is only one among countless star systems within the universe.

The Earth is solid nickel and iron at the core, surrounded by a molten outer layer. Mixtures of many other elements, including radioactive ones, make up the mantle and crust.

Early Astronomers

Anaximander	610-546 B.C.	Introduced the use of the sun dial as a time-keeper. Pictured the sky as a sphere with the Earth floating in space in the center.
Heraklides of Pontus	388-315 B.C.	Explained that the apparent motion of the stars was a result of the Earth's rotation. Believed that Mercury and Venus revolved around the sun, but that the Earth was the center of the universe.
Hipparchus	146-127 B.C.	Devised the geocentric theory--that the sun and planets revolve upon an epicycle, the center of which revolves around the Earth.
Nicolaus Copernicus	1473-1543	Discovered that the sun was at the center of the solar system, which accounted for the apparent motion of the planets in the sky.
Tycho Brahe	1546-1601	Made very accurate measurements of the positions of stars and the movements of planets.
Johannes Kepler	1571-1630	Developed the first three laws of planetary motion.
Hans Lipershy	1587-1619	An optician; invented the telescope.
Galileo Galilei	1564-1642	Discovered the phases of Venus, identified four of Jupiter's moons, discovered sun spots and mountains on the moon.

Until the latter part of the last century, it was erroneously believed that the Earth had originally been a ball of fire. This idea seemed to account for the fact that the irregularities on the surface of the Earth had come about as a result of shrinking and wrinkling during a long period of gradual cooling. Ruth Moore, in her book *The Earth We Live On*, describes how in 1885 Thomas C. Chamberlin, a noted geologist who later became president of the University of Wisconsin, changed that theory by proposing an entirely new hypothesis. Chamberlin proposed that there had indeed been an original ball of fire, but it had been much smaller than the size of the present Earth. He proposed that those portions of the Earth which constituted the mantle and the crust arrived here from outer space; the Earth had been brought to its present size by gradual accumulation of meteorites very early in Earth's history during a period of accretion. This theory is given further credence by the finding on the surface of the moon many impact craters where meteorites have struck with great force. Ancient metereorite craters on the Earth have been erased over time by the effects of erosion. There are meteorite craters on the Earth; however, they are fairly recent.

What were these materials which had gathered around the core? Scientists now know that these were stony meteorites and particles which had been floating in space. These were created when stars burned out in spectacular supernovas. Thus it was that Earth received the supply of elements necessary for the development of man.

Aerial view of a metereorite crater located in Canyon Diablo, Arizona.

16 In 1806, John Lightfoot proposed that Earth came into existence in the year 3928 B.C. In 1630, James Ussher, an Anglican archbishop, calculated from biblical data that the Earth was created in the year 4004 B.C. His belief that this was the date of creation was so strong that he had a statement to that effect included in the Bibles of the day. Both of these calculations approximate the date of creation in the Jewish tradition, which likewise is based on biblical records. (The Jewish calendar is 3760 years ahead of the Christian calendar; thus 1983, for instance, is 5743 in the Jewish calendar.)

Until the past few decades there was no means of determining the age of the Earth. It is now possible to fix this date with a high degree of accuracy. Scientists have found one of the important threads of the tapestry and have traced it through from the reverse to the proper side. This thread was the discovery that there is, within rocks, minerals, metals, and meteorites, a mechanism by which the time of their origins can be accurately determined, since certain elements decay into other elements. These are the radioactive elements, sometimes called radioactive isotopes. Radioactivity can be measured quantitatively in terms of the half-life of the element. For example, uranium decays into several forms of radioactive lead. Strontium decays into thorium, and radioactive potassium decays into the gas argon. By meticulously measuring the rates at which uranium, strontium, and potassium have decayed, scientists found that all three measures indicate that our planet came into existence 4.6 billion years ago. At this time elements within meteorites came together to form the mantle and crust of planet Earth.

How large is the Earth? The average diameter of Earth is 7,900 miles. The radius at the equator is 3,963 miles, whereas from the center of the Earth to the poles, where the Earth is somewhat flattened, the radius is 3,953 miles.

Looking southward from the north pole, it can be observed that the Earth is rotating in a counterclockwise direction, causing the sun to appear to rise on the eastern horizon and to set on the western horizon. For some unknown reason, the planet Venus, which is somewhat smaller than planet Earth and only 66 million miles away from the sun, rotates in the opposite direction and requires many Earth-days for the completion of one rotation. Due to the rotation of the Earth, the equatorial areas of the Earth receive sunlight for an average of 12 hours of each 24 hour period. The Earth also rotates about the sun, which requires approximately 365 days, giving us our calendar year. The time required for a complete rotation of the Earth about the sun is slightly more than 365 days. So each four years, in the leap year, a day is added in the month of February.

The Earth's pathway around the sun is not a circle. Any two-body system produces an elliptical orbit, because each exerts a gravitational attraction on the other. When the two bodies are of such disparate size as the Earth and the sun, one rotates about the other. Although the Earth rotates around the sun, the sun also has to compensate for the pull of the Earth, and there is a tiny perturbation in its own revolution. Johannes Kepler in the 17th century was the first to describe these elliptical motions mathematically.

Still another aspect of the Earth is its polar tilt. The orbits of the planets are also affected by the gravitational pull of the rest of the solar system. The north pole tilts toward the sun for about six months and tilts away from the sun the other six months. When the sun's rays are at a greater angle, the northern hemisphere experiences fall and winter; when the sun's rays are more direct, it experiences spring and summer. The opposite is true in the southern hemisphere. The tilt is maximum at what we know as the summer and the winter solstices. Thus the four seasons are produced.

The equator was not always where it is now. About 300 million years ago, when a great proliferation of living forms occurred, the equator coursed through Montana, British Columbia, Prudhoe Bay, and into Siberia. Fossils of tropical plants and dinosaurs have been found along this route.

What causes the tremendous irregularities on the Earth's surface – the depths of the oceans, the high mountain peaks, the deep gorges, the plains, and the valleys?

The crust of the Earth is made up of what are known as tectonic plates, which ride on a plastic mantle. When these plates shift or overlap, a phenomenon known as an earthquake occurs. There are certain areas on the Earth's surface where so-called faults are present; these faults are cracks or divisions between plates. Such areas are more prone to the occurrences of earthquakes than others. The western coast of North and South America, from Alaska to California, to the west coast of Mexico, Guatemala, and the rest of Central America are particularly susceptible, as are the islands and coasts of East Asia and Southeast Asia. The entire rim of the Pacific basin has come to be known as the Ring of Fire. In earthquakes as in volcanic eruptions, the surface of the Earth is literally torn asunder. Mountain areas come into being by the impact of land masses colliding with each other via the movement of the tectonic plates.

The surface of the Earth is greatly affected by erosion, the gradual wearing away of the Earth's surface by wind and water. Erosion breaks down rocks, creates riverbeds and canyons, and levels mountains. It creates sand dunes, caves, mesas and plains. Erosion also creates sedimentary deposits, layers of material laid down through wind and water, which over time are transformed into rock. The advance and the retreat of the great ice masses, another form of erosion, have also had their effect on the contours of the Earth's surface. Glaciers level mountains, and also scoop out lakes. They form hills of rock and dirt deposits.

18 Without volcanic action, life could never have evolved. There would have been no atmosphere, no water, and no assembly of life's necessary components. In the center of the Earth, surrounding its core, there is no fusing hydrogen as in the sun, but there is still, and always has been, a great cauldron of magma. The stony meteorites which came to the Earth at the time of accretion were melted and transformed by tremendous heat into a semi-liquid, so that eventually the entire mantle of the Earth assumed some degree of plasticity or fluidity. As the materials closest to the core underwent physical and chemical changes due to superheating, gases and water vapors were produced, just as they are in the kiln of the potter. In a kiln, as the clay becomes vitriolized and both physical and chemical changes occur, water is produced. If the kiln is not properperly ventilated, a tremendous explosion will occur. In the center of the Earth, at the molten core, similar processes went on early after the formation of the Earth. This still occurs today in the extreme upper mantle. Under the influences of tremendous heat, water vapor and other gases which have been created expand and produce tremendous pressure. When the pressure exceeds the strength of the mantle and the crust, the magma rises and a volcano is formed.

First, there is an immense cloud of gases – carbon dioxide, hydrogen sulfide, and perhaps some ammonia. Repeated explosions follow, thrusting lava and flaming boulders high into the sky. Your author will never forget the spectacle of the baby volcano called Paracutin in the neighborhood of Uruapan, Mexico. He stood knee-deep in ash watching in the darkness, hearing the rumbling – and then the explosion, the thrust into the sky of flaming boulder masses, and their fall back to the lava cone where they rolled along with the fiery lava flow to its base. This process went on for months.

In all probability, thousands upon thousands of volcanic actions occurred since the beginning of Earth's history, and they are still occurring in the 20th century. Without them we could never have been. Volcanos contributed the first atmosphere, though it contained no free oxygen. Without volcanos there would have been no water, for when water vapors condensed, they produced rain. As rain flowed down the sides of volcanos it formed trickles, creeks, streams, and rivers which eventually pooled to form oceans. Eroded materials were carried along with the flowing waters to produce sediment; some was suspended, some precipitated, and some dissolved to form the saltiness of the sea.

When Earth was in a partially molten state, it acquired a satellite which we call the moon. Just as the Earth revolves about the sun, the moon revolves about the Earth, requiring 28 days for each journey. For a portion of the month, when the moon is full or near full, it provides light at night. Whether waxing or waning, however, it exerts a gravitational pull upon our great bodies of water, thereby creating the tides. Is it mere coincidence that dogs howl at the moon, that the emotions are disturbed by a full moon, that in women the menstrual periods occur every 28 days? Why should grunions come in with the tide and plant their fertilized eggs where they will remain to develop in the warm sand until another full moon and another tide washes them back into the sea to develop and to eventually repeat this urge? And, there are those who say that plants do better when seeded in the light of the full moon. But the light of the moon is only reflected light from the rays of the sun.

The movement of the Earth as represented by its rotation on its axis, the pursuit of its elliptical course around the sun, polar tilt, and other minor motions of the Earth, are all better understood it they can be related to their beneficial effects on all life. These motions keep the Earth's relationship to the sun intact. Without the sun, nothing on Earth could live. Photosynthesis could not occur, and without plants the entire food chain would be destroyed. Vitamin D could not be elaborated and the bones of the members of the animal kingdom would be soft, pliable, and nonsupportive.

Spacecrafts Viking I and II carried with them to Mars sophisticated little black boxes and probing arms which could be extended to pick up samples of soil. These samples were tunneled into compartments within the little black boxes where the soil was analyzed for sign of life. Nothing could be proven conclusively from the samples, but there is probably no life, as we know it, existing on Mars.

Probings of the distant planets by the spacecraft Voyager indicate that conditions on both Neptune and Saturn are so extreme that life as we understand it could not long survive. Some scientists speculate, however, that there may be life on Jupiter's moon Europa, for underneath the thick layer of ice which blankets the planet there may exist liquid water. As far as the scientists know, all the other planets are too near or too distant from the sun, making them too cold or too hot, and their atmospheres, for the most part, prohibit life. Free water has not been found on any planet other than Earth. Our planet was provided with everything conducive to the creation of and the survival of living matter. Insignificant and small as it may be when compared to the other components of the universe, planet Earth, nonetheless, has distinctive attributes. It is alive with potential.

It is like the composer who has within his grasp the basic elements and is ready to unite them into themes, counter-themes, and variations on themes that will enter into a master composition. And so it is with the Earth. Our group of travelers notice that, after 1 million years of existence, our planet is still a sterile thing. But it has composed itself; it has superbly accomplished the first part of the symphony, that part which will lead out and eventually reach the crescendo. And what is this crescendo? It is the creation of life itself, its evolution.

Part 2: LIFE COMES TO PLANET EARTH

Chapter 5: CONCEPTS OF LIFE

"Life cannot be truly analyzed; it can only be meditated upon and alluded to in terms of the divine."

Rabbi Manfred Swarsensky, Temple Beth El, Madison, Wisconsin, 1981

Our group of travelers now view an Earth which is one billion years of age. Matter is present in its three fundamental forms: solid, liquid, and gas. The primary liquid of the Earth, water, is composed of two atoms of hydrogen in combination with one atom of oxygen. The solids of the Earth, making up land and the bottoms of the streams and oceans, are numerous and varied. The atmosphere surrounding the Earth contains carbon dioxide (one atom of carbon linked to two atoms of oxygen), nitrogen (two atoms of nitrogen), and hydrogen sulfide (two atoms of hydrogen linked to one atom of sulphur). Some solids are dissolved in the waters of the sea and some are suspended by the buoyancy of these waters. The atoms are there with their electrical charges, subatomic particles, and the tremendous amount of energy contained within each. Along with inorganic compounds which have been fashioned by the combination of certain atoms or groups of atoms, all are waiting to give their individual qualities in response to an ageless evolutionary urge.

What were the next steps in this steady progression? At one billion years of age, the Earth was much hotter than it is today. Since there was no free oxygen in the air, there could be no ozone layer to filter out the ultraviolet rays of the sun. Along with carbon dioxide, a major part of the atmosphere, this created a condition known as the "greenhouse effect." The waters of the streams, seas, and shallows were maintained at a high temperature. Elements and compounds were dissolved or suspended in the waters of the seas, streams, and the shallow ponds which had been warmed by the heat of the sun. This created an ideal situation for chemical reaction and interaction. This warm primordial ooze (or soup, as it was called by J.B.S. Haldane) formed the ideal locus for chemical reactions which might have led to the beginnings of life.

These atoms, compounds and molecules undergo repeated interaction, repeated combinations, and must be repeatedly subjected to the play of various natural forces upon them. The chemical laboratory contained within these warmed waters had within it many potentials, many possible sequences which might have led to an entirely different evolutionary trend, an entirely different quality of life, or to some other quality which was not life. However, there was disregard of those processes and sequences which were not adequately tuned to the evolutionary symphonic composition of the future. The ancestral past of planet Earth demanded that, even though the Earth was growing more complex, the same orderly progression which had governed the past would be strictly obeyed. Thus it was that out of the confusion of the primordial soup there was established this one, proper, perfect sequence, the only direct and irreversible progression which could lead to the beginnings of life as we know it.

But life is not all that simple to define. Life's chemicals and their various combinations, structures and forms are well known indeed. What about that spark, that something beyond our comprehension which causes these combinations of chemicals, structures, and forms to function? Without the function, the same ingredients, structures and forms would be inchoate, and Earth would be little better off than it had been at the end of its first billion years of existence. This spark, called "life," will constitute a factor over which our mythical travelers must ponder for the remaining portion of their journey, 3.6 billion years into the 20th century future. They must become aware that the unique quality of life cannot be resolved in definitive terms. Only certain aspects can be made to respond to our current technologies, and then only in fractionated terms.

Other aspects of life are so profoundly mysterious that they are beyond the realm of human comprehension. We can make only minimal and partial analyses in an attempt to gain some understanding of that mysterious quality, called "life."

The many facets of life can, usefully, be divided into three categories: first, all that is knowable, then all that is unknowable, and finally a gray area constituting a symbolic bridge (but a bridge that cannot quite be crossed) between the knowable and the unknowable.

The knowable aspects of life – consisting of atoms, molecules, compounds, and various combinations thereof and of structure and form, and some of the physical forces which react upon them – have been subjected to all sorts of scientific probings and searchings. As a result, these facets of life have become more and more completely known. There is great expectancy that even more knowledge will accumulate. The unknowable, however, lies beyond any scientific approach.

Throughout the ages the most erudite minds have devoted their thoughts to metaphysical philosophies, casting about for any threads which might lead into that mysterious realm of the unknowable. The most profound scholars and thinkers have failed to cross this bridge. Yet the knowable and the unknowable have accompanied each other throughout all of cosmic time and space. As our mythical travelers go forward they will see continuing interactions, not only to create life from the inanimate, but to cause it to continue to evolve through the most magnificent complexities. Ultimately, it will cause these chemistries, structures, and forms to possess function – for indeed, function is living.

Scientists know that living matter can adapt to its environment. It can absorb certain ingredients from its surroundings, metabolize them, and produce products which are necessary for its growth and maintenance. It can excrete its by-products and, most amazingly, can reproduce its own kind. But scientists do not know the nature of the spark which excites these combinations of chemistry, structure, and forms to initiate and maintain those functions which constitute the differences between the inanimate and the animate.

In the 1840s there lived in Hungary a young lad by the name of Ignatz Philip Semelweis. He lived in Buda across the Danube River from Pescht. He was later to become famous for his discovery of the cause of the great killer of childbearing women of the time, an overwhelming infection known as puerperal fever. He devised a simple preventative means: having the attendants thoroughly wash their hands in chlorine water, thereby preventing infectious material from being introduced into the birth canal. As a schoolboy in Buda, Semelweis was very curious, and one day while attending a lecture in biology he stood, faced his professor, and questioned as follows: "With your permission, Herr Professor, sir, why can a bird fly?" The professor, much annoyed, answered, "Semelweis, everyone knows that a bird can fly because it has feathers." Young Semelweis was not quite satisfied with this answer and he again stood, faced his professor, and addressed him as follows, "With your permission again, Herr Professor, sir, last evening on my way home from school I found a bird. It was lying in the street. It was motionless. I picked it up. It had feathers, and yet it could not fly."

This anecdote illustrates, in a simple way, the futility of trying to define life in its totality. The beautiful thing that Semelweis held in his hands had been a creature long evolved from amphibians, reptiles, dinosaurs, and from the archaeopteryx, the first living thing capable of flight. It was composed of all of the same things as all other birds – the chemistry, the structure, and the form – yet it could not fly. It could not reproduce; it could not sing.

There was something lacking. This "something" which we cannot deny is necessary and ever present in all that lives. We can allude to it only as something beyond our comprehension. Without it there was nothing in the motionless bird to ignite a spark in the central nervous system, causing the signal to be passed along nerve pathways to the many structures which, functioning in a coordinated way, produce life. This stilled creature had lost this mysterious something which could never be regained. And life is indeed like that. It begins with the simplest ingredients of the universe, yet it culminates in structure and form which are necessary for that mysterious function called life which prevails for a given span and then departs, leaving what is inanimate to disintegrate, to be recycled and returned to become a part of that vastness from which it had originated.

But what about these components of life which are not within the realm of the tangible, visible, or knowable? Do they also become disintegrated, disassembled, and disavowed? Or do they return to that part of the cosmos which is unknowable, and become forever contained therein as immortal?

In the environs of Spring Green, Wisconsin, lies Taliesen East, the home of the late famous architect, Frank Lloyd Wright. Near his self-designed home, there is a small cemetery in which the immediate members of his family are interred. On one of the tombstones there is engraved the following: "The soul belongs to eternity. No matter how long it occupies the body it never grows old." This concept of the soul as synonymous with the essence or spirit has throughout all of recorded history been the center of man's deepest meditations. The essence has even been thought to have had issue with a divine creator. Most of the great religions hold that when the processes of life cease to exist, the body disintegrates and returns to that from which it came, but the soul becomes immortal and lives forever.

Over 5,000 years ago in ancient Mesopotamia, in the lands of Babylon, Sumeria, and Assyria, men wrote such thoughts on tablets of baked clay. These beliefs have not been altered in all of recorded history. Many scientists and theologians do not accept these beliefs, however. They adhere so strictly to their individual beliefs that they leave no room for dialogue. They fail to realize that the knowable and the unknowable have always been a part of one and the same thing. As long as the two groups resent each other's intrusions into their specific belief commitments, there will be no common understanding.

There are barriers to the mystical approach of the theologians as well as to the materialistic searchings of the scientists. Dr. Robert Jastrow puts it very well in his book titled, *God and the Astronomers:* "At this moment, it seems that science will never be able to raise the curtain on the mystery of creation. For the scientist who has lived by his faith in the power of reason, the story ends like a bad dream. He has scaled the mountains of ignorance. He is about to conquer the highest peak; but, as he pulls himself over the final rock, he is greeted by a band of theologians who have been sitting there for centuries."

There is no place for the Inquisition, for punishment, persecution, or death against those who have stated their views about things in which they truthfully and conscientiously believe. For both the creationists and the mechanists, the road ends before the complete story is unveiled. There will always be an unknown. Goethe put it well: "The shudder of awe is man's highest faculty." The great thinker, Spinoza, was a Spanish Jew who had fled to The Hague in Holland. His thoughts were contrary to the theological teachings of the day. In the year 1677, in great secrecy, he published his famous treatise entitled, "Tractatus Theologico Politicus," in which he defines the thoughts which prevailed concerning the nature of creation and of all living things. He advocated complete freedom of thought and expression about all things with regard to philosophy and religion. He espoused pantheism, believing that all things in nature were identical with God and that all things were a part of one infinite substance. These thoughts are not too far removed from some of the themes of these writings. Nor are they far from the mythologies of ancient Mesopotamia, which relate that all within the realms of the inanimate and animate are part of all that ever was, is now, or ever shall be. Life and non-life are intimately bound together, and that which is easily accessible to our knowledge cannot be separated from that which is not. Therefore, let us deal with that which is within our power to grasp, and ponder reverently about that which is beyond our comprehension.

An engraving showing the type of torture that Galileo feared. During the Inquisition, those accused of heresy risked torture or death by the Holy Office.

Thornton Wilder has said, "Out of man's recognition in fear and awe that there is an unknowable comes all that is best in the exploration of the mind." Despite limitations and regardless of barriers which are placed before people, they will continue the quest, trying to get answers to that which seemingly lies beyond their grasp. Nothing should ever deter them from seeking answers concerning the ultimate metaphysical reality. But the full appreciation of the richness of all that is will, in all probability, never be obtained in all of time. Perhaps it is best that it is this way. If all knowledge were complete, every facet of the magnificence of creation revealed, there would be no reason to continue to search and probe, and the need to use and develop the mind would be diminished.

With these thoughts in mind, our travelers, with profound humility, proceed with exploration into the future, where they will encounter wonder upon wonder, far beyond anything which could possibly have existed in their most imaginative thoughts. As we accompany them, may we, too be humbled by what will unfold before us.

Part 2: LIFE COMES TO PLANET EARTH

Chapter 6: THE BROTH OF GENESIS

Before living matter could come into being, it was necessary that the place, the time, and the ingredients be brought together in nature's laboratory, a place where all the factors were primed for this great adventure. We have now seen how this broth of genesis was assembled. The elements were in outer space when stars at the end of their life cycles consumed themselves as novae and supernovae. The elements formed the Earth as they were attracted by the force of gravity. The hot core of Earth became superheated and molten.

Gases, including water vapor, were generated. They expanded under the intense heat, creating tremendous pressures. Great explosions extruded the gases, and lava and great volcanic cones were built. The gases supplied an atmosphere. Condensation of the water vapor produced rain, finally accumulating to form great oceans. Sediments from water- and wind-driven erosion were carried along by rushing waters and poured into the sea providing salinity and an accumulation of atomic ingredients. The warm, shallow waters – absorbing gases of the atmosphere, teeming with atomic elements which formed molecules, and galvanized by the sun and by lightning – became nature's laboratory, charged with the challenge of producing living matter. This, then, was the broth of genesis.

Planet Earth arrived at its one billionth birthday, and after this long period of preparation the broth was ready. Let us examine the contents. What were

Lightning may have been the energy source which contributed to the creation, within the primordial soup, of the first organic molecules on Earth. Photograph shows the birth of Surtsey Island off the coast of Iceland. Steam and other gases are being emitted from the volcano.

the specific ingredients, what were the forces which caused them to interact with each other, and what would be the final result?

In the year 1805 John Dalton determined that all matter was composed of units called atoms. There are many types of these basic units, each having a specific weight. The lightest and most ubiquitous of these atoms is hydrogen, and the heaviest naturally occurring one is uranium. These basic units of matter were much later classified according to their weights in what came to be known as the atomic periodic table. But Dalton did not identify all of the natural elements. Furthermore, there are now also a number of manmade elements heavier than uranium. All of these are unstable, some lasting only a fraction of a second in the laboratory before decaying into another radioactive element. Others have half-lives of many thousands of years; for instance, the half-life of plutonium is 24,000 years. Plutonium is perhaps the best known of these manmade transuranic elements. Computers predict that the atomic scale can go as high as 115 elements. Assume, then, that in this broth, made from many of the original 92 atomic elements, were liquids, solids, and absorbed gases. Assume, also, that the shallow waters were warmed and lighted, speeding up the reactions of the chemicals. This indeed was nature's laboratory at its best.

We recall that the atoms of each element carry one or more electrons, which, interacting with the atoms of other elements, can bind them together to form a molecule. For example, a carbon atom can combine with four atoms of hydrogen, creating a molecule of the compound CH_4, commonly known as methane or marsh gas. A carbon atom can combine with only two atoms of oxygen resulting in carbon dioxide (CO_2). As one further example, a molecule of water (H_2O) is formed of two hydrogen atoms and one oxygen atom. These few examples illustrate the principle. Thus, it can be understood how the broth of genesis became an exciting medium.

Within this broth the potential was vast. Thousands upon thousands of reactions were possible. Many of them took place but led nowhere. One series of reactions and interactions, one series of sequences, however, did lead somewhere. It led to the formation of methane and then to the first organic molecules, the amino acids, the building stones for the more elaborate protein molecules without which living matter could not exist.

Part 2: LIFE COMES TO PLANET EARTH

Chapter 7: THE CHEMISTRY OF LIFE

An ancient Chinese proverb of unknown source states, "Truth goes like the river, never in a straight line. There are curves and bends, shallows and deeps, narrows and wide spreads, each facet of which must be explored when searching for the ultimate truths." Dr. Herbert Clark, Emeritus Professor of Botany at the University of Wisconsin-Madison, believes that certain chemosynthetic bacteria preceded all other living forms. In November 1977, a team of scientists working at Northern Illinois University (Carl Woese, Ralph Wolfe, and George Fox), announced their discovery of a bacteria-like organism which had been thought to be perfectly ordinary, but which proved to be a separate form of life. This organism may represent the oldest living thing on Earth. It can survive in an atmosphere devoid of oxygen and at temperatures of 170° F or greater. It is able to live in hot, thermal springs in swampy marshes where decaying materials have used up all of the oxygen. It also lives in the intestinal tract of certain animals, where likewise, there is little or no oxygen.

The genetic material in the ribosomes of this bacteria-like organism is distinctly different from that found in any other living bacteria. This organism is capable of reproducing its kind by the process of binary fission, a simple division. It can absorb from its environment carbon dioxide and hydrogen, metabolize them, and excrete methane (marsh gas) as a by-product. Methane has proven to be a necessary link in the chemical sequences leading to the formation of amino acids and subsequently to the formation of protein molecules. Its importance to these sequences cannot be over-emphasized.

Dr. Richard E. Dickerson, in an article titled, "Chemical Evolution and the Origin of Life," (*Scientific American*, September, 1978) theorizes that the organism discovered by Woese, Wolf, and Fox might have originated from non-biological materials during the first billion years of the Earth's existence. Some ancient stony and metallic meteorites have been discovered in the Fig Tree area of South Africa. By means of radiometric techniques, the ages of these meteorites have been determined to be 3.2 to 3.4 billion years. Within these meteorites were fossils of microorganisms which would have had to survive and replicate and function under the environmental conditions which existed at the time – that is, with no free atmospheric oxygen and with temperatures of 170 degrees Farenheit or greater.

These bacterial fossil forms could have been identical with the methane-producing bacteria discovered by Woese, Wolfe, and Fox – who have shown that the ribosomal genetic materials in this organism are distinctly different from those of all other known living forms. They speculate that these bacteria could have been the source from which all forms of life's evolutionary development proceeded.

These same bacteria, which originated when the Earth was approximately one billion years old, can still be found in the 20th century, under environmental conditions such as those that prevailed when they originated.

Now that methane has been added to those ingredients previously present, what could be the next step in the orderly series of sequences that lead to the development of living form and structure? Ruth Moore, in her book *The Earth We Live On*, reports the interesting research procedures devised by Dr. Stanley L. Miller and Dr. Harold Urey. They created an artificial atmosphere consisting only of carbon dioxide, water, methane, ammonia, and hydrogen sulfide, similar to that which existed billions of years ago. Miller and Urey contained this atmosphere within a flask and subjected it, in one instance, to ultraviolet light, and in the other to an electric spark similar to that which might have been created by lightning. In both instances, daily examinations of the contents of the flasks revealed the presence of several amino acids. However, many more amino acids were present in the contents of the flask which had been subjected to the electric spark than in the flask subjected to ultraviolet light.

Thus, the stage was set for the elaboration of the protein molecule. Dr. George S. Mileaux, professor of Geology at the University of Strasbourg, states that many investigators suspect clays may have played a crucial part in the origin of life. Several researchers tested such theories by simulating conditions

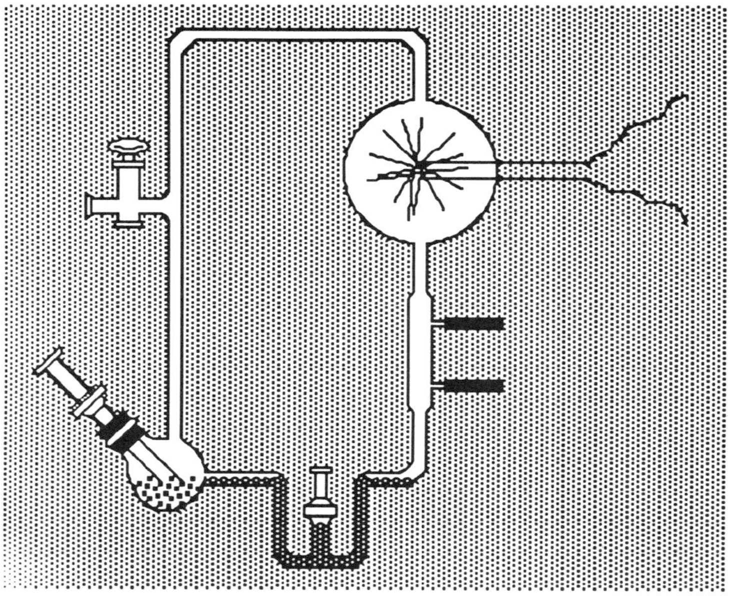

Schematic diagram of Miller-Urey experiment, described above.

which might have existed prior to the beginning of life. Among these researchers were Dr. David White of Santa Clara University and Dr. Sherwood Chang of the Ames Research Division of the National Aeronautics and Space Administration who worked with a simple amino acid called glycine. This was placed in a watery solution containing various clays and the solution was subjected to different temperatures. When the solution was dehydrated and again rehydrated, long chains of amino acids had linked to form peptides and polypeptides similar to protein molecules.

Later, when this solution was again dehydrated and rehydrated, apparently a cleansing action of the clay particles occurred. New complexities of amino acids resulted, leading to more elaborate protein molecules. It was demonstrated that the amino acids were, indeed, the building stones for the protein molecules. By the utilization of 20 amino acids, all of the protein molecular components of living matter can be produced.

The stage was set for acceptance of the entire concept of molecular biology; but, what about nucleic acids, the other component so necessary to life? Deoxyribonucleic acid (DNA) and ribonucleic acid (RNA) are responsible for carrying and executing the entire genetic code which is involved in the reproductive processes of all living matter.

Four of these nucleic acid bases (adenine, guanine, cytosine, and thymine), together with a phosphate and a sugar called a ribose, combine to form ribonucleic acid. DNA differs from RNA in that thymine replaces uracil and the deoxy form of ribose is present. As the theme of these chapters develops, the importance of these two groups of chemically compounded substances – amino acids and nucleic acids – will be emphasized. They are absolutely necessary ingredients of every living structure and form. Without them life as we know it could not have possibly come into existence. DNA and RNA will likewise become a part of every basic unit of life, for without them no perpetuation of living species could occur. Without them there would be no place for the genes to reside; there would be no heredity, no diversification; all of life would be confined to one monotonous simple form. But it could not be that way, for it has been designed otherwise.

Part 2: LIFE COMES TO PLANET EARTH

Chapter 8: THE STRUCTURAL UNIT OF LIVING MATTER

Here we are with our mythical travelers, standing before a warm and lighted shallow pond on a 2 billion year old planet. Here have been assembled all of the components necessary for the structure of living matter. And while the onlookers have gazed at these waters, within this broth of genesis the elaboration of methane, amino acids, and nucleic acids has been completed.

Before these amino acids and nucleic acids could fulfill their most important tasks in the cycles of life, it was necessary that they be contained within a structure or form. To meet this need, within this same broth of genesis, other chemistries were at work. The fatty or lipoproteins came into existence and formed globules, and in some way the DNA and RNA were surrounded by this globule of fatty proteins. Such globules were the precursors of cells.

These protocells could live in an atmosphere devoid of free oxygen and where the temperatures were much higher than those which would be encountered later. But, it is necessary to make one point clear: these primitive cells functioned as living beings, for they absorbed nutrients, they changed them, and – as in the case of the very first bacteria – they produced and excreted methane. They had within themselves the units necessary for procreation, the molecules of DNA, molecular structures which they would share with all future cells. There were perhaps several species of these so-called chemosynthetic bacteria out of which a tremendous refinement of the first living things was to evolve.

As we stand at the edge of this warm and lighted shallow pool, something is happening which had not been observed before. Floating on the pond's surface and adhering to a partially submerged stone there are patches of a frothy, bluish-green substance from which emerge bubbles of gas. What is this substance, and why does it bubble?

Full answers to these questions have not been easy to come by. It was not until the 17th century A.D. that what was happening in this pond became fully understood. Within these blue and green pigmented masses were myriads of microscopic, spherical structures which were the prototypes of all future cells. These were distinctly different from those first bacterial cells which could live only in high temperatures and in the absence of free oxygen. They were a new type of cell which could survive and reproduce at lower temperatures, and in the presence of oxygen. These were the original blue-green algae, not quite plant and not quite animal, but the forerunner of both. They possessed the unique ability to photosynthesize – a process which was to change the whole course of evolution.

Beneath the outside membrane of each of these cells some metallic blue-green pigments had assembled. This material, now called chlorophyll, could either trap or filter out certain light rays from the sun. The cells could absorb carbon dioxide from the atmosphere and from water. Within the cell, the filtered light could disassociate the atoms of carbon dioxide and of water, rearrange the electrons, and produce new forms, sugar and starch, and store them for use in maintenance and growth.

The by-product of this chemistry was oxygen. Some of the available oxygen was used for the metabolic processes of the cell itself for maintenance, growth, and reproduction. But much of it bubbled out from the cell. Thus, for the first time, when planet Earth was about 2 billion years old, its atmosphere contained free oxygen. The full significance of this particular cellular function will become apparent.

Detailed knowledge of cell structure had to await the discovery of a means of magnification, for the eye alone could not see individual cells. Just as all non-biological and biological structures advanced from the simplest to the more complex, so it was with the technology which eventually made it possible to augment man's visual acuity to the degree where it became possible to penetrate the secrets of the cell. The spectacle-makers first came upon the scene. They ground and polished pieces of glass so that the two sides were convex, but to different degrees. Such glasses enabled those with dimmed vision to see better. The principle of magnification did not come into use until Galileo constructed a low-power telescope by means of which he could see ships coming up over the horizon before they could be perceived by the naked eye. When he turned this newly found instrument toward the vastness of the sky, he learned not only that the moon had mountains but that the planet Jupiter had four moons.

In Galileo's day people were intellectually restless. They searched for something more than mere platitudes to explain the universe which surrounded them. It was therefore not strange that searches began to extend into microscopic phenomena as well as into the macroscopic.

In the mid-17th century the British scientist Robert Hooke was well aware of these principles of magnification as they had been employed by the spectacle makers and by Galileo. He began to utilize such ground pieces of glass as a new tool to study the details of everything which came within his grasp. New worlds opened up before him.

On one occasion, Hooke placed a thin slice of cork beneath such a magnifying glass and, for the first time, the eyes of man were able to view the structural units of living matter. Hooke saw that this piece of cork was composed of compartmentalized units uniform in size and adjacent to each other. Each was completely separated from the one contiguous to it. They seemed to be like minuscule rooms so, appropriately, he called them "cells." In the year 1665, Robert Hooke published his observations in a book titled *Micrografica*, and the world of cellular biology was on its way.

Not long after, a Dutch janitor by the name of Anton van Leeuwenhoek had been grinding a lens after the fashion of spectacle makers, and out of curiosity, he placed a drop of pond water under such a lens. There he discovered that everywhere within this droplet were swimming, darting, whirling little creatures the like of which had never been seen before.

Throughout the years to follow, this new tool, the principle of magnification, was to be used in more and more sophisticated instruments. Even the most minute detail of structure, form, and function would come to be a part of general knowledge.

In the year 1824, the Frenchman H.J. Dutrochet made extensive microscopic examinations of many living plant and animal tissues, and described their cell-like structures. It was not until 1838 and 1839, however, that two German scientists, M.J. Schleiden and Theodore Schwann, published the results of their extensive studies on the microscopic structures of plant and animal tissues. From then on, the concept of the cell as the basic unit of all living plant and animal tissues was firmly established. Schleiden and Schwann affirmed that each cell was surrounded by a limiting membrane in which was contained a fluid which they named cytoplasm. Within this fluid there was a globular structure which they designated as the nucleus. They further stated that a living cell could give rise to a new living cell. The limiting membrane underwent constriction, then separation occurred, and the limiting membrane grew over the wound caused by the separation. The extruded cell grew, so that it achieved the dimensions of the cell from which it had come. They called this process "free cell formation." Repeated examinations have verified the original findings of Schleiden and Schwann. The principles of cell structure, function, and procreation which were elucidated by these pioneer scientists have become well established, and form the basis for our continuing 20th century research.

Present-day scientists have many advantages over these pioneers. Related scientific technologies and increased power of magnification have come

Upper)- Division of a human cell showing chromosomes seen here as long, dark strands.
Lower)- Division of an onion cell . Chromosomes separate to form two new cells.

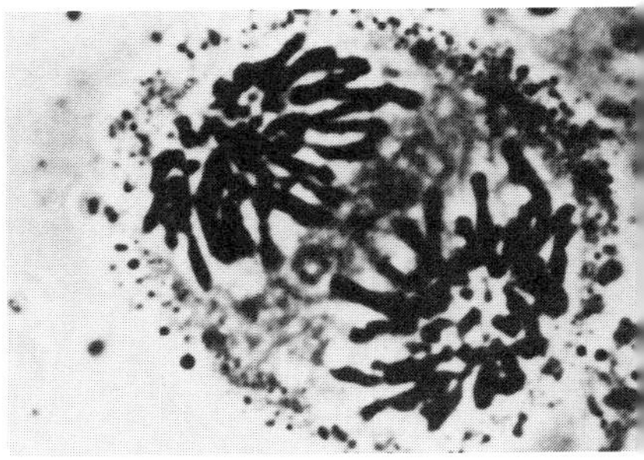

 about from development of electron and scanning microscopes capable of almost unbelievable magnifications. As an additional aid in cellular studies, pigmented stains selectively color various intracellular structures to which individual functions can be assigned.

Composed of elemental atoms from outer space, compounded and elaborated to perform the most sophisticated functions, the broth of genesis provided the means by which living matter could become manifest. But living matter is perishable. It is born, it lives and functions for a time, it procreates its kind. When its usefulness is finished it dies, to disintegrate and return to the elemental forms from which it came, perhaps to be recycled into other living structures and forms.

Thus it is with the planet Earth and with the living matter which so abundantly occupies its surface. This is a great symbiotic relationship. Both the Earth and we, ourselves, have a common origin, and our interworking, our interrelationships, have been present for a long, long time. They will continue until our source of energy from the center of our solar system, the sun, exhausts itself. For it is certain that without warmth, without light, without electromagnetic forces, and without the gravitational attraction of our sun, living things could not possibly continue to be created nor to survive on planet Earth. Such a catastrophe seems distant, for our scientists predict that the sun contains adequate fuel to produce energy for several billion years. Perhaps other disruptive forces will destroy life before the sun loses its life sustaining beneficence.

Part 2: LIFE COMES TO PLANET EARTH

Chapter 9: THE ANCESTRAL ALGAE

At this time, our limited knowledge might suggest that, among all of the trillion or more planets of the universe, the planet Earth is unique in possessing living matter. The validity of a statement such as this is open to controversy, because in this, the latter part of the 20th century, the exploration of other planets its just beginning.

Manned spacecraft have visited the moon. Our astronomers found the moon to be a sterile thing, without atmosphere and without moisture, and, therefore, without life.

Unmanned Viking I and II have landed on Mars. The sophisticated little black boxes on the Viking landers were designed by Dr. Carl Sagan, Professor of Planetary Sciences at Cornell University, and by Dr. Joshua Lederberg, Nobel Prize winner formerly of the Department of Genetics at the University of Wisconsin-Madison and now President of Rockefeller University. The purpose of these boxes was to discover, if possible, any evidence of life on the planet Mars. No life was found, at least not as we know it.

Venus, the next planet to come under investigation, is slightly smaller than the Earth, but only 63 million miles from the sun compared to 93 million miles for Earth. Probings by the two Explorer spacecraft revealed that Venus is surrounded by a deep layer of carbon dioxide and hydrogen sulfide, and the temperature is over 900° C – conditions which earthly forms of living matter could not tolerate.

Jupiter and Saturn, whose orbits are farther removed from the sun, have climatic conditions too severe for the existence of living forms. Saturn is about 238° F at its surface. Pluto, even farther removed, is in all probability considerably colder.

So far, our explorations have extended to relatively near planets within our own solar system. How about other suns and their planets in the Milky Way? Our galaxy is 120,000 light years in diameter. That is a distance almost too great to comprehend. Yet, within this space there are hundreds of billions of stars, each with its own planetary system. With this great number of planets, can it be possible that another planet like Earth might possess all of the essentials for the creation and maintenance of life? With all of the possibilities, it seems likely that out there, somewhere, there may be another planet Earth. Multiply the number of stars in our galaxy by the billion other galaxies, and the possibilities seem more likely. The distances are so great that if life did or does exist and messages were or are sent, many human generations would pass before an answer could be received.

One bit of evidence may cause us to be suspicious that, in outer space somewhere, life may be present. Certain meteorites have found their way from outer space to planet Earth. One meteorite found in Australia and several others found embedded in the blue ice of Antarctica contain amino acids. Life on planet Earth began with organic compounds such as these.

Nobel Prize winning scientist Frederick Crick, who with James Watson first described the structure of the DNA molecule, has theorized on the principle of "transpermia," or nature's way of transplanting organic material from one planetary system to another. This theory, however, seems highly improbable, for it is generally conceded that meteorites come into being from the matter of stars which have consumed themselves to extinction. The possible discovery of life in outer space most probably must await the development of technology which will permit instruments to travel to these far, far distant places.

Our little world is very important at this moment in the journey of our travelers. At this point, Earth is two billion years of age. It has prepared itself for the advent of life, has created that first primitive form of life, the chemosynthetic bacteria, and then the blue-green algae. These blue-green algae could not only create oxygen by the process of photosynthesis, but they could utilize it, excrete it, and live in its presence. The entire future of evolutionary biology changed, for the atmosphere contained free oxygen, which was used in the metabolic processes of these tiny little creatures, providing energy for growth, maintenance, and reproduction of their kind.

Robert Dott and Roger Batten, in their book *The Evolution of the Earth*, wrote as follows: "South African algae and bacteria are the oldest known living things, whereas Southern Rhodesia contains the oldest and largest reef-like masses, about 3.5 billion years of age." These calcareous reefs have been gradually and slowly built up by algae. In agreement with Dott's and Batten's figures, fossils of the methane-producing bacteria in South Africa, similar to those which had been researched by Woese, Wolfe, and Fox, have been similarly dated.It is thought that the chemosynthetic bacteria slightly predated the coming into being of the algae.

Dr. J. Wm. Schopf, Professor of Paleontology at the University of California, Los Angeles, in an article titled "The Evolution of the Cell" (*Scientific American*, September 1978), reported that these discoveries were of tremendous importance in any historical study of the evolution of life. Precambrian sediment was laid down about 600 million years ago, and until the discovery of bacterial and algal fossil remains in South Africa and the reefs off the coast of Rhodesia, it was thought that this was the time at which life had begun.

Prior to these discoveries the only fossils were of such organisms as shellfish, trilobites, and worms. It was not until the last three decades of this century that evidence of microscopic fossils began to accumulate, and they could be accurately dated only with the development of radiometric techniques.

In the early part of this century, Charles Doolittle Wolcott, while exploring in the Pacific Northwest for specimens in Precambrian sediments, came upon some strange-appearing rock formations. He described them as looking like one pancake piled upon another, several layers deep. He called them stromatolites.

Typical forms taken by stromatolites which contain fossil algae.

The late Dr. Stanley Tyler, research professor in the Department of Geology at the University of Wisconsin, studied the microscopic bacteria associated with metal-containing ore. In the year 1957, when Tyler was searching for specimens in the Gunflint area along the northern shores of Lake Superior, there came into view some strange, crystallized rock formations. Dr. Tyler's widow recounts that these rocks were in a rather inaccessible place and on several occasions Dr. Tyler had walked away from them. However, his curiosity kept bringing him back and eventually he succeeded in obtaining the specimens and putting them into his bag. Upon returning to the University, he placed them on a shelf in the laboratory, where they rested for several weeks before he got around to cutting them into thin slices for microscopic study. Using the technique of the oil immersion lens, which clarifies detail of the object being studied, he learned that these were stromatolites built up of fossilized algae cells. He immediately called his wife by telephone and in great ecstasy declared that he thought he had found the fossils of the oldest living cells on Earth. He thereupon consulted with Dr. Barghoorn, paleobiologist at Harvard University, who confirmed that these, indeed, were the fossils of ancient forms of algae. They were similar in all respects to those found by Charles Walcott, and those which had been discovered on the reefs of Rhodesia and dated as 3.5 billion years old. Since then, stromatolites have been found at some 45 or more sites, in widely scattered areas. All have contained fossil forms of algae, indicating the widespread proliferation of this cellular structure.

This dispersion of algae, together with other plant sources of free oxygen, brought the oxygen of the atmosphere from zero to 20 percent, our present level.

Bruce Parker and George Semmons, of Virginia Polytechnical Institute and State University, while exploring the bottoms of Lake Hoard and Lake Frixell in Antarctica, discovered some unusual pinkish-orange pigmented algae growing in mat form in water just above the freezing point. Under these conditions, it was possible for them to receive only one percent of the available light. Yet this pinkish-orange pigment permitted algae cells to maximize this small amount of light for the photosynthetic process. Finding such algae cells in the remote Antarctic is remarkable because in these regions there are four months of total darkness, four months of twilight, and four months of continuous daylight.

Algae diversified and spread to all parts of the Earth. In addition they adapted to extremes of climate and to various amounts of available light. Algae cells so changed the atmosphere that all cellular structures of the future would have to survive under new environmental conditions. The free oxygen in the atmosphere created an ozone layer which began to filter out some of the ultraviolet rays of the sun. Without this layer, we would all suffer from ultra-violet burning. With time, there occurred a gradual lowering of the Earth's surface temperature to the present range.

Our mythical travelers are impressed that one little cell type could affect the entire course of evolution. With three and one-half billion years to go, this is still only the beginning.

Part 2: LIFE COMES TO PLANET EARTH

Chapter 10: PLASMIDS, VIRUSES AND CELLS

Earth, air, fire, and water – these are the ingredients of which life is made, said the ancients. In a sense they were correct, for earth, air, and water are composed of the atomic elements, often in the form of molecules. The majority of these elements are contained within the solid components of the Earth. The waters of the Earth are made up of one simple molecule consisting of two atoms of hydrogen and one of oxygen in which many of the solids are either dissolved or suspended. The air (or the atmosphere), which was belched up from the interior of the Earth by volcanic action, prior to the creation of the algae cells was comprised of only carbon dioxide, hydrogen sulfide, and perhaps some nitrogen. All of these ingredients contained within the air, within the Earth, and within the water, in all probability would have remained without living forms were it not for another component, fire.

Most certainly there was within the center of the Earth extreme heat, which helped to bring to the surface many of the atomic elements and compounds essential to the creation of life. Again, we want to stress that life could not have been, were it not for the heat and light provided by our sun through the fusion of hydrogen nuclei. The energy from this fusion reaction is transferred to the surface of planet Earth by electromagnetic radiation.

Biologists know that even the simplest, most primitive bacterium possesses a cell wall which encloses a fluid cytoplasm, DNA, RNA, and proteins. A cell is able to absorb certain chemical ingredients from its environment. It is able to metabolize them and to excrete certain by-products. It can grow and maintain itself; and, very importantly, it can reproduce its kind.

As evolution of cells has advanced, there have been more acquisitions to internal structure and function, leading to formation of more complex cells. But even as this progression has occurred, basic components have remained. Take away one – soil, air, water, or heat – and life as we know it could not exist. On a global basis, this is not likely to happen because these ingredients are ever present. The Earth is surrounded by atmosphere; water covers approximately 75 percent of the Earth's surface; and the sun continues to bestow its beneficence upon the land and the sea. Water composes as much as 99 percent of some cells; but, in others, non-water molecules may weigh as much as 4,000 times the weight of water molecules. Some cells can become dormant if the amount of water within them is materially reduced. Just as in the broth of genesis, the ongoing chemistry within a cell is best performed within this watery medium. With warmth and light the cell's performance is enhanced.

Were the primitive bacteria, the methane-producing bacteria of Woese, Wolfe, and Fox, the earliest and the simplest forms of life? Had there been other non-cellular living entities which had preceded them?

Discoveries within the last few decades indicate that there might have been such non-cellular life forms. These are the plasmids and the viruses, neither of which has the structure of the cell. Plasmids and viruses are so ultramicroscopic that no fossilized forms of them have been discovered, and no radiometric dating can be obtained. But we shall deal with their structures and potential functions alone, and learn, if we can, in just which category they should be placed.

Could the plasmids and the viruses have been the mere chemistries necessary to the ultimate creation of the cell, or were they the first forms of life prior to the creation of cells? Did they exist prior to the time when planet Earth was one billion years of age?

Dr. Richard P. Novick, of the Public Health Research Institute of the City of New York, considers plasmids to be subcellular entities, "poised on the threshold of life." Plasmids are quite simple in structure and function compared to the viruses. They are found within the cellular structures of bacteria. They are nothing more than circular molecules of double-stranded DNA. They have no limiting membranes as have viruses and cells. They live and function within the bacteria cells where they can reproduce and be passed on to the progeny of the bacteria. They provide a means of accessory inheritance other than that provided by the DNA of the bacteria.

Should plasmids be lost, bacteria are not capable of reproducing them. They can be acquired only from other bacteria. When bacteria mate, genetic information is carried from one bacterium to another. Dr. Edward Tatum, biochemist at the University of Wisconsin and later Professor of Biology at Stanford University and the Rockefeller Institute, and Dr. Joshua Lederberg, Professor of Genetics at the University of Wisconsin, were intrigued by the plasmid transfer of certain nutritional needs from one bacterium to another. They won the Nobel Prize for discovering that bacteria can pass on their hereditary qualities. Plasmids can also confer upon the cell structure the means to resist certain toxic and even antibiotic effects. Dr. Novick states that they have introduced non-bacterial genes into the bacteria, thereby providing a set of new,

Electron micrographs of three types of primitive bacteria. From upper to lower: Escherichia coli, found in human intestines; Spirillum volutans, found in fresh water ponds and streams; and Streptococcus lactis, found in milk.

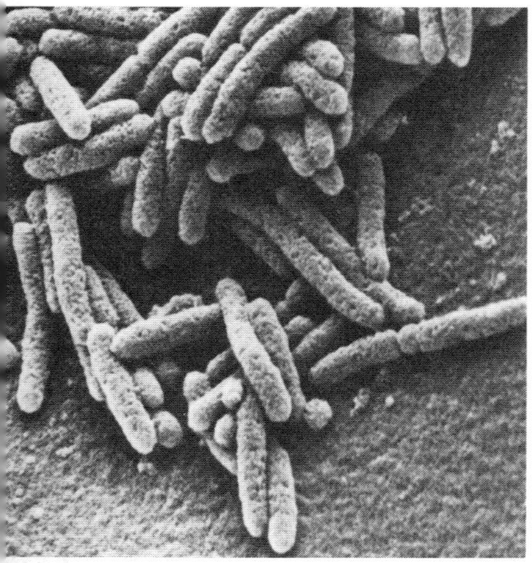

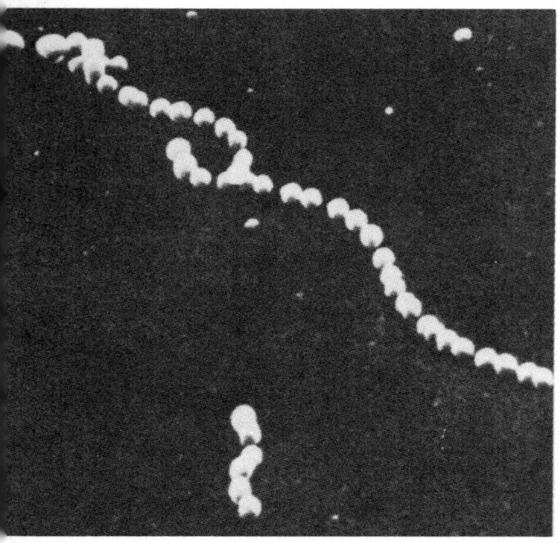

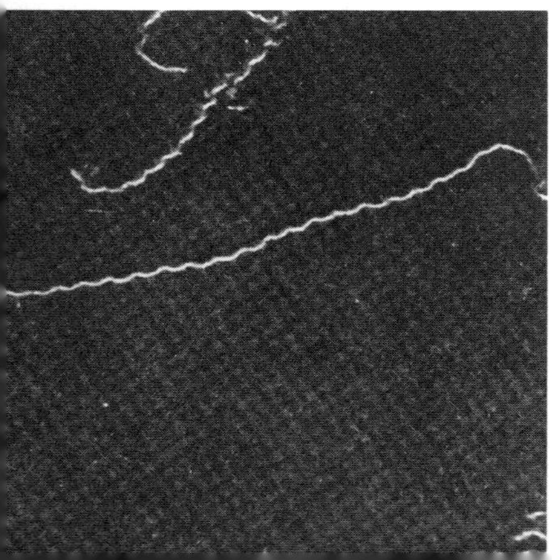

hereditary qualities.

Some of these genes can produce proteins which are distinctively different from those which the bacteria could produce. This is of profound interest to medicine because it may be a means of producing insulin for the treatment of diabetes and, perhaps, interferon for the treatment of infectious diseases and the experimental treatment of cancer.

There is more and more evidence that plasmids are non-cellular forms of life. Independently of the cell, they create copies of themselves, and they provide accessory inheritance qualities for the cell itself. This is perhaps the earliest form of a symbiotic relationship, about which more will be learned in the next chapter while dealing with the eukaryotic or true cell.

Plasmids, the earliest form of the chemical combination of purines or pyrimidines, sugar, and a phosphate capable of participating in the hereditary processes, deserve more, it seems, than to be classified along with non-living matter.

Viruses are one step beyond plasmids, but have not quite reached the status of cellular forms of life. They consist of a protein and a piece of nucleic acid, either DNA or RNA. They may be contained within an envelope. They can survive, but cannot replicate or reproduce, outside of cellular structure. They can penetrate the cell membrane, find their way to the nucleus, and reproduce themselves. In the process, they can destroy the cell.

Geneticists have found viruses to be valuable tools for the introduction of new hereditary qualities into the cell itself. They are responsible for producing many serious, sometimes fatal diseases; but they also are a means of producing immune substances for protection against disease.

From what has been learned in recent years about plasmids and viruses and their relationships to early cellular forms, it seems about time that our concepts of their role in evolutionary biology be revised. Perhaps they are transitional forms between inanimate and animate. They contributed much to the creation of what previously was considered to be the first forms of life - the chemosynthetic bacteria.

Upper) - Electron micrograph of a bacteriophage (virus), showing tail fibers. Lower) - Electron micrograph of bacteriophages attacking a cell of E. coli. Note that the viruses are attached to the cell by their tails. Empty-headed viruses have already injected their DNA into the cell.

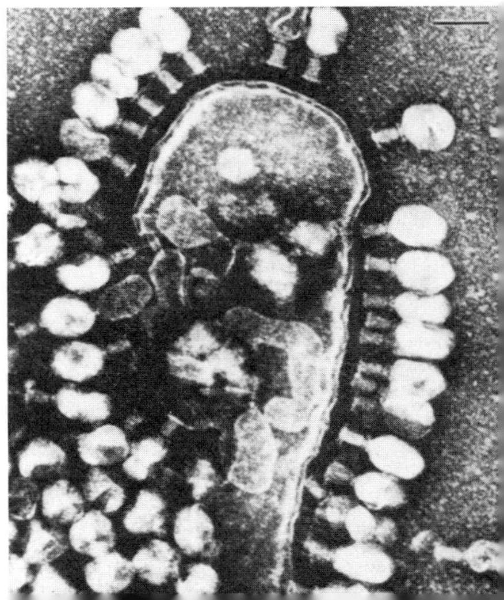

Part 2: LIFE COMES TO PLANET EARTH

Chapter 11: LIVING FORM AND FUNCTION GROW MORE COMPLEX

Our mythical travelers have learned that by the time Earth reached its one billionth birthday, the plasmids, viruses, and a few cellular bacterial forms had already been created. These living forms belong to a very special class, in that they cannot tolerate free oxygen, and their DNA and RNA are not enclosed within a central nucleus. Now, our travelers are about to come upon an entirely new type of living cell which, typically, cannot exist without free oxygen in its environment. Its genetic mechanism has become more elaborate and is contained within the cellular structure of the nucleus.

This new cell is known as the eukaryotic or true cell. In order to place this new cellular type in proper evolutionary progression, it seems quite necessary that your author provide some repetition and elaboration and clarify, for the reader, some of the important evolutionary sequences, and how and why they occurred.

We return for a moment to the most sophisticated of these earliest cellular structures – the blue-green algae. These were the cells which had adapted and diversified to the greatest extent. Having evolved new structures, they could now produce a whole new series of products, and thereby change the course of evolutionary history. The prokaryotic blue-green algal cell is a microscopic rounded form. The wall which surrounds the inner ingredients is made up of fatty proteins. As seen under the microscope, this wall is not a simple flattened surface but is convoluted like the cortex of the brain, which gives it greater surface exposure for absorption and excretion.

Directly beneath this limiting membrane of the cell are accumulations of bluish-green pigment forming the photosynthetic lamellae, precursors to the chloroplast. The rest of the cell consists of its fluid part, called cytoplasm, and fragments of DNA. These structures permit the algae cells to perform a function which earlier cells had not yet achieved.

Carbon dioxide and water can now, in this blue-green algal cell, be absorbed through the cell wall into the cytoplasm. After this has been accomplished, the miracle begins. The bluish-green pigment of the photosynthetic lamellae have the ability to select out certain light rays from the sun. This disassociates carbon dioxide into carbon and oxygen, and water into hydrogen and oxygen, and then recombines the carbon, hydrogen, and the oxygen to produce sugars and starches which are stored in the cell for later use. However, more oxygen is produced than the algal cell needs. This surplus oxygen is excreted back into the atmosphere, the original source of the atmospheric oxygen now essential to life. A few types of bacteria were able to adapt to oxygen in the atmosphere, able to live with either its presence or its absence. But to most bacterial forms shortly following the first one billion years of Earth's existence, oxygen was toxic.

Algae cells, however, not only found it possible to generate oxygen, but also to live in its presence. All new true cells (or eukaryotic cells) find atmospheric oxygen absolutely necessary for their survival. This is indeed a most important evolutionary turning point. These fully developed, nucleated algae cells have adapted and diversified and have come to occupy almost every niche of planet Earth's surface. They are to be found in the waters of the oceans, lakes, and streams, in and upon the soil, on both moist and arid surfaces, in all temperature extremes, and even on particles in the atmosphere.

It is not known how many varieties of the algal cell there are. But imagine the amazement of the Dutch janitor Anton van Leeuwenhoek when in the year 1702 he looked at a drop of pond water under a piece of ground glass. There, for the first time, a human being was confronted with a new world of swimming, darting, whirling, living, microscopic beings impossible to see with the naked eye.

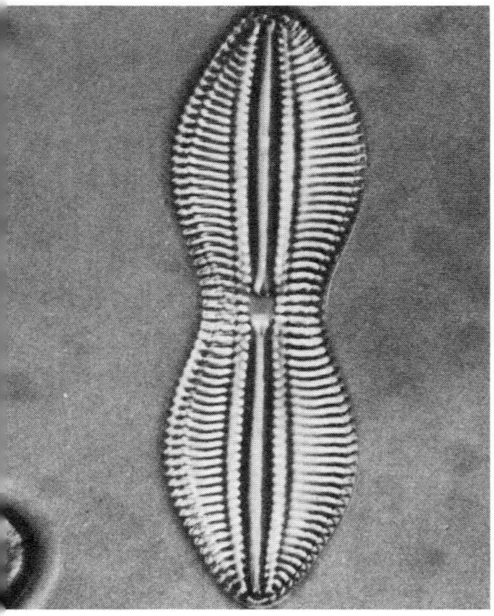

A diatom showing its intricately marked shell. The fine lines represent passageways connecting the interior of the cell with the environment.

These microbes, or "little animals" (as they were described by Leeuwenhoek), were the diatoms. Their name is Greek for "split in half." This designation, in all probability, had been on the basis of observing that when the algal cell reproduced it seemed simply to split itself in half, making two instead of one. It is now known that this process is not one of simple binary fission.

Diatoms have adapted and diversified to the extent that, at present, there are known to be over 25,000 species. Ten million of these little creatures can be contained in one liter of sea water. Algae such as these, and the various primitive bacterial forms, are sometimes referred to by the biologist as simple forms of living matter. "Simple indeed, are they?" it might be asked. They seem simple in form and in structure. When the multiplicity of function and the benefits which have accrued from them are brought into focus, however, the term "simple" seems totally inadequate. These minute creatures constitute an important part of our ancestral past. Fortunately many of them are still with us in this, our 20th century.

Marine plankton (their name is taken from the Greek word "planktos," meaning "wandering") include both plant forms (the phytoplankton, including the diatoms) and animal forms (zooplankton). They are all single-celled creatures. The zooplankton, the animal cells, consume the diatoms and other algae, and thus the food chain is started. They are food sources for larger and larger creatures, up to and including gigantic whales. If the algae had never come into being, there would have been no food chain, or at least it would have been quite different.

The diatom has other unique qualities which benefit human beings. For example, it can extract from its marine environment the soluble silicon, change it into the salt form, silicate, and deposit it around its soft body, creating an outer skeleton for protection against predators. Later, this practice was employed by shellfish. When the diatom dies it sinks to the bottom of the sea with billions and billions of others of its kind, sometimes forming deep layers of stony material known as diatomite. In the process, the fatty material within the soft inner tissues is squeezed out and, over eons of time, collects in large reservoirs. These are now the pools of fossil fuel so necessary to satisfy our modern energy needs. The diatomite, when powdered, has such fine texture that it has been useful in industry. Diatomaceous earth is a perfect filter for certain chemical processes, and is also used as an abrasive for conveying a high polish to some metal surfaces. It is easy to understand the importance of these tiny, not so simple, miraculous little creatures. How well these very early forms of living matter have extended their usefulness into modern time!

Dr. Richard B. Hover, radio astronomer by profession but diatomist by hobby, has beautifully described these many benefits in an article titled "The Miraculous Myriad Diatoms," in the June 1979 issue of *National Geographic*. He has studied diatoms throughout the world. He describes how Gerald Mueller, a German scientist, mounted on a single microscope slide over 4,000 of the species of diatoms known at the time. While visiting in Antwerp, Belgium, where he had gone to examine a collection of diatoms, Dr. Hover was given a piece of blotting paper on which some diatoms had been dried in the year 1834. Yet, when this piece of blotting paper was placed in water, it was not long before the diatoms had regained their watery contents and became alive again as if nothing had ever happened to them.

In the September 1978 issue of *Scientific American*, Dr. J. William Schopf, paleobiologist at the University of California-Los Angeles, published an article titled "The Evolution of the Cell." In this article, he fully documents the earliest findings of the fossil remains of the eukaryotic or true cells, the ones containing a central nucleus. These cells were somewhat larger than the primitive non-nucleated cells.

The earliest of these fossils were radiometrically dated at 1.4 billion years. Therefore, it seems to have taken over 2 billion years for these nucleated cells to evolve. What had these primitive algae cells been up to during this vast period of time? The true cell requires about 20% oxygen in its environment for its optimal life processes. Perhaps it had taken that long for a sufficient number of algae cells to be created in order to bring the oxygen content of the atmosphere to one in every five molecules (20%). It might have been, also, that these primitive bacteria, including the algae, joined with their new, evolutionary neighbors, those true single-celled creatures, in a communal type of living.

Algae and chemosynthetic bacteria joined with this new type of nucleated cell. This combination enhanced their efficiency, bestowed special benefits on all three, and at the same time preserved their individuality. A single algal cell found its way through the wall of the nucleated cell and into the cell cytoplasm. There it denuded itself of its limiting membrane, and arranged its substances along the inner wall of the nucleated cell. This substance of the algae included the pigmented chlorophyll, its own DNA, and its own cytoplasm. It then became an organelle, a working unit of the nucleated cell, and became known as a chloroplast. It had, in its new location, the same capabilities for photosynthesis as it had in its original state.

At the same time, some of the chemosynthetic bacteria found their way into the cytoplasm of the nucleated cell. They carried with them their own cellular compounds, including DNA. These became known as mitochondria. As residents of the new cell, they could call forth energy from the starches and the sugars of the nucleated cell's cytoplasm and convert it into energy, or adenosine triphosphate. By this means, energy release increased by about 33 percent over the amount released by the process of fermentation which they had used when they were only chemosynthetic bacteria.

In the mid 1920s, when your author was a student of medicine at the University of Wisconsin-Madison, one of his professors, Dr. Theodore Bast, was teaching histology, or micro-anatomy, as it was then called. The scene is still very vivid: as this white-coated gentleman peered through the microscope at a tissue preparation, he exclaimed with great excitement, "Those little threads are the mitochondria. It is not known why they are there or what purpose they serve, because they have just been discovered. But you must look, for there they are." Since that time, much has been discovered about the origin of the rod-like structures and about the role they play in the economy of the cell.

We can visualize the symbiotic relationship going on within this nucleated cell. First, there is the fluid known as the cytoplasm, enclosed within a limiting membrane. In one portion of the cytoplasm is the nucleus, also enveloped by an enclosing membrane. Within this nucleus there are chromosomes, which contain DNA molecules in double strands wound about each other in helical fashion. All of the biological molecular processes of the cell are controlled in this particular area of the cell.

In a portion of a plant cell is the chloroplast, the representation of the algal cell. It retains its capabilities of photosynthesis, producing energy which is stored in sugars, starches, and fatty materials.

In an animal cell there are organelles called mitochondria, which represent the original bacterial forms of life. The mitochondria convert nutrients into usable energy. During cell division, chloroplasts and mitochondria, simultaneously but independently, are called forth by their DNA molecules to reproduce, also. This triumvirate, this community relationship, this symbiotic association is truly a most remarkable series of circumstances. It is even more so because

A diagram of an animal cell. The cell is bound by a membrane, which is selectively permeable so that materials may enter or leave the cell.

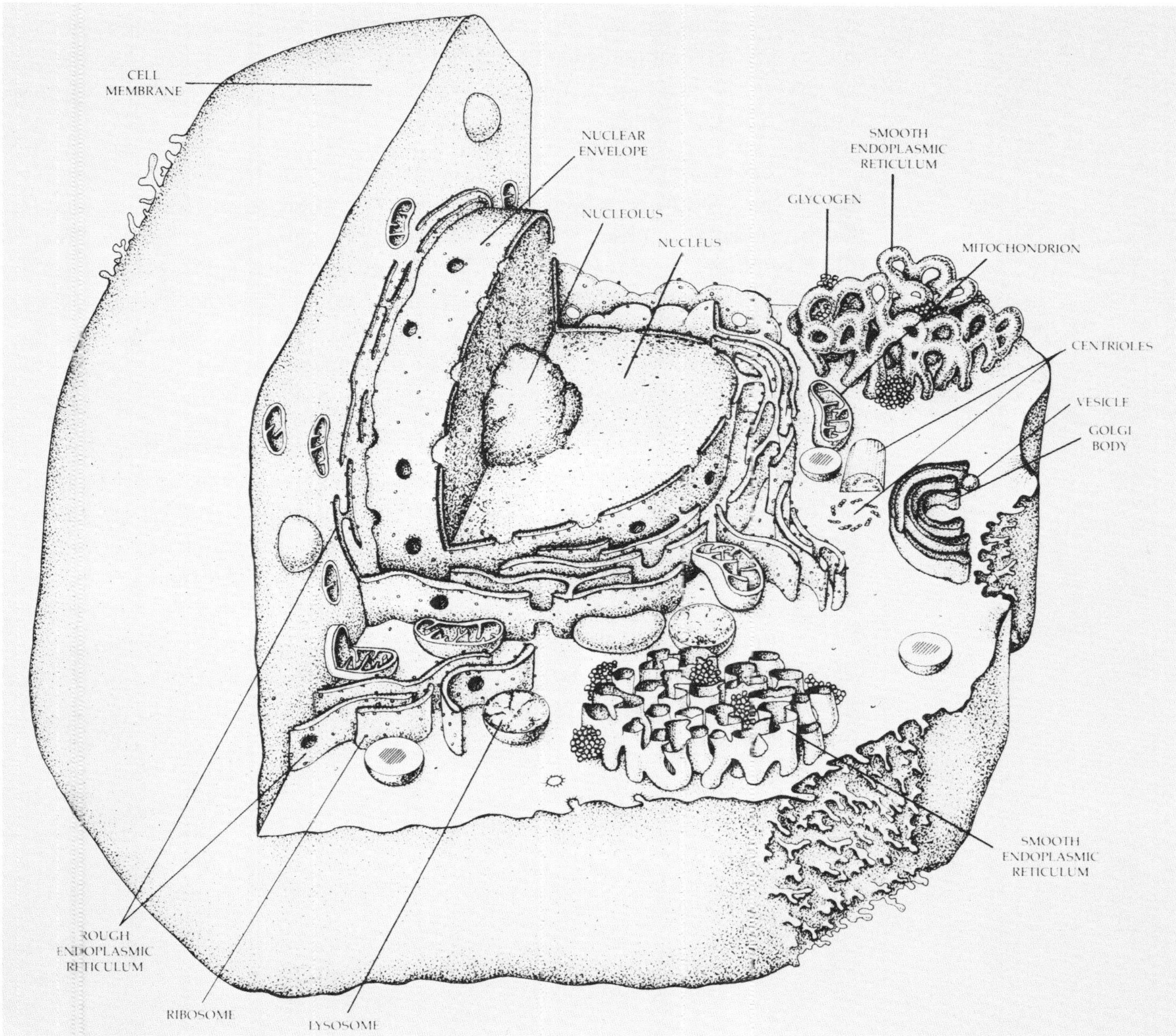

all nucleated cells, both plant and animal, have either chloroplasts or mitochondria, respectively, within their cytoplasm. All nucleated cells will continue to pass these characteristics on to their progeny.

This comprises what has come to be known as symbiotic heredity. It is this three-part heredity combined within one cellular structure which at one time wrinkled the brows of many biologists who considered this idea to be highly controversial. However, this is no longer so. Dr. Larison Cudmore in her book *The Center of Life* clarifies this subject extremely well. Dr. Lewis Thomas, President of the Sloan Kettering Memorial Cancer Research Institute, in his book *The Lives of a Cell* expresses confirmatory views. Likewise, Dr. Lynn Margulis in her book *The Origin of the Eukaryotic Cell* has documented the evidence so well that little doubt can be entertained as to the existence of these symbiotic relationships. Several other geneticists with whom your author has had personal communications regarding the views expressed here have indicated that they, too, have strong personal leanings towards acceptance of these unique facets of evolutionary biology.

Sequence upon sequence has led from the simplest to the more complex organisms. With these complexities more and more refinements of function have resulted. Nucleated cells of 1.4 billion years ago have incorporated within them symbiotic relationships. Symbiotic heredity will continue into the future as long as the nucleated cell remains a part of the evolutionary process. These views are in perfect harmony with the theme of these writings: that we, as human beings, participate in all of biological phenomena in which our remote, single-celled ancestors were involved.

Have these first cellular forms of life, bacteria and algae, together with the first of the nucleated cells been no more distinctive than to deserve classification as simple life forms? The designation of "simple" could in no way describe the multiplicity of their beneficial attributes. How and why have these progressions come about? If they were accidental and happenstance, then creationists have something on which to build their claims. If present complexities of structure, form, and function of biological entities have resulted from a series of progressive sequences building one upon another, each more nearly perfect than those which have preceded, then the theory of evolution appears to be more tenable.

Part 2: L FE COMES TO PLANET EARTH

Chapter 12: BIOLOGICAL PERPETUATION

> The fairest thing we can experience is the mysterious. It is the fundamental emotion which stands at the cradle of true art and true science. He who knows it not and can no longer wonder, no longer feel amazement, is as good as dead, a snuffed-out candle. It was the experience of mystery – even if mixed with fear – that engendered religion. A knowledge of the existence of something we cannot penetrate, of the manifestations of the profoundest reason and the most radiant beauty, which are only accessible to our reason in their most elementary forms – it is this knowledge and this emotion that constitute the truly religious attitude; in this sense, and in this alone, I am a deeply religious man.
>
> Albert Einsten
>
> *The World As I See It*

It took about 3.2 billion years from the beginning of planet Earth's existence for the true nucleated cell to come into being. Our mythical travelers are poised for the last 1.4 billion years of their journey. They have seen, in addition to bacteria and algae, the true cell develop to become the basic structural unit of all members of both the plant and animal kingdoms. Some plants and some animals are made up of a single cell only, while others have from a few to an enormous number of cells. In the latter, such cells often are differentiated into organs, and the organs work in unison to produce a living whole.

Some of the inner workings of the true cell have been discussed in the preceding chapter; however, there is one subcellular part, the nucleus, which needs further exploration.

The nucleus has many important roles to play in the functioning of the cell. Within the cytoplasm we see it as a rounded mass that stains strongly when the tissue is prepared for microscopic study.

Cells which make up living tissues have many things in common. They come into existence, they live for a given span of time, and then they die, disintegrate, and return to the non-living materials which gave them origin. If this were all, Earth, in time, would return to the sterile thing that it was. However, within the nucleus of cells there is a magnificent mechanism to ensure perpetuation of their kinds. In addition, there is provided a means for variations to occur in structure, in form, and in function. Certain cells (such as red blood cells) have no nucleus, and of course cannot reproduce themselves.

Much of what is known concerning the nucleus has been discovered only within the last few decades. The nucleus is the molecular headquarters for control and for direction of the functions of the cell, for its reproduction and for the perpetuation of its hereditary characteristics. These are huge tasks for one small part of a cell to accomplish, and yet it is now known that this is where it all takes place.

There are three molecules which are of great importance to cellular reproduction and to cellular heredity. They are the molecules of DNA (deoxyribonucleic acid), RNA (ribonucleic acid), and protein. DNA, which is found in the nucleus, carries instructions for the production of RNA and protein and directs the functions and reproduction of the cell.

Scientists have long been aware that both males and females were necessary for animal and human reproduction. They also know that offspring frequently do not resemble their parents.

In the middle of the 19th century, Gregor Mendel, an Austrian, joined the Augustinian order of monks in his home town of Brunn, Austria (now Brno, Czechoslovakia). In the monastery gardens, he began to experiment with peas, crossing plants with certain characteristics with others that did not possess those characteristics. In all, there were seven pairs of contrasting characteristics including wrinkled vs. smooth seeds, short plants vs. long plants, and yellow vs. green seeds.

When plants grown from wrinkled seeds were pollinated by plants from smooth peas, the pods contained not a single wrinkled pea. When Mendel allowed plants from these smooth peas to self-fertilize, they produced about three times as many smooth as wrinkled peas – a ratio of 3:1. When plants from wrinkled peas were self-fertilized they produced only wrinkled peas, but plants from smooth peas either produced only smooth peas or again produced a 3:1 ratio of smooth to wrinkled. Mendel explained these results by assuming that each pea has a pair of determining factors, W and w, for smooth and wrinkled. The original smooth strain was WW, the wrinkled was ww. The first generation hybrids were Ww, having gotten a single factor from each parent. Because the W factor is dominant to the recessive w, these peas were smooth. When these hybrid plants were self-fertilized, random combination of the factors led to 1/4 WW, 2/4 Ww, and 1/4 ww. The first two classes were smooth, thus giving the 3:1 ratio that Mendel found. Plants from WW peas, when self-fertilized, produced only smooth seeds, while those from Ww plants produced a 3:1 ratio of smooth to wrinkled.

These principles held true with all of the other characteristics that Mendel studied. He was thereby able to establish that there is some direction and control over heredity, and that the rules follow simple mathematical principles. The units of heredity, now called genes, are present twice in each individual, and one of the two is transmitted to each offspring. Usually, but not invariably, one member of a pair is dominant to the other.

These principles came to be known as the Mendelian laws of heredity. His writings on this subject were little noted at the time, however, even though, in the first half of this century, there were many practical applications of these principles (such as cross-pollination of vegetable and flowering plants). The unraveling of the threads of this tapestry began about 30 years ago when scientists, with new thoughts and new technologies, worked unceasingly to solve the riddle of cellular reproduction and heredity. They began to dissect the nucleus, which had long been known to be in control of the reproduction of hereditary characteristics.

The period from 1900 to 1930 was one of rapid understanding of the mechanism of heredity. It had been known since the latter part of the nineteenth century that the thread-like bodies visible in the nucleus, the chromosomes, undergo a precise separation at every cell division so that the number remains constant in all descendant cells. It was also known that the chromosome number is halved in the production of sperm and egg cells and the double number is restored at fertilization. Thus it was quickly surmised, and later proved, that Mendel's genes are carried by the chromosomes. Furthermore, the genes are arranged in a definite order along the thread-like chromosome. But, despite the deep knowledge of where the genes are and how they are and how they are transmitted from one generation to the next, nothing was known of their actual natures.

Prior to the early 1940s, scientists thought that simple organisms such as bacteria invariably reproduced themselves by simply splitting into two identical halves. This process came to be known as binary fission. If this assumption were true, then each and every cell resulting from such a process would have been absolutely identical to that from which it had come, and biological evolution could not have progressed.

Well into the 1940s it had been accepted that bacteria, although they clearly had inherited traits and therefore genes, did not have chromosomes.

Diagram shows the result of Mendel's experiment in crossing a smooth seeded pea plant with a wrinkled seeded pea plant.

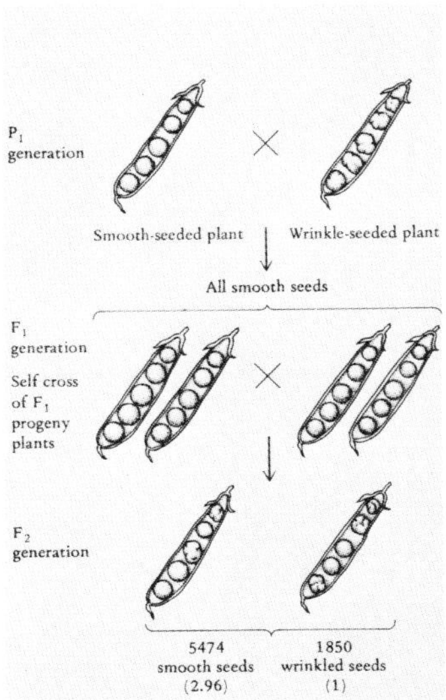

 There was no mating and therefore no mixture of genes from different cells. Then there came upon the scene two brilliant scientists, Dr. Edward Tatum, Professor of Microbiology at Yale University, and Joshua Lederberg, his student. They observed that when two strains of bacteria containing contrasting genetic traits were grown in the same culture medium, they produced colonies of cells not found in either of the two strains. There had been some sort of sexual process by which the genetic traits of the two strains were scrambled. This opened up the possiblity of doing genetic experiments in bacteria. Later, other workers discovered that genes can also be transferred from one bacterium to another by viruses.

The work of Lederberg and Tatum was based on the earlier work of Tatum and George W. Beadle, who had worked together at Stanford. They succeeded in showing that, in fungi, genes are responsible for the synthesis of vitamins and amino acids. Strains lacking these normal genes could not grow unless the appropriate vitamin or amino acid was supplied by the experimenter. Lederberg and Tatum used the same kinds of genes to show sexual reproduction in bacteria. They reasoned that if nutritional characteristics could be inherited, so could other characteristics. This was a breakthrough, the result of a research program well designed and meticulously carried out. It showed that the propagation and the variations of all living creatures are similar, and are dependent upon and controlled by these nuclear entities, or genes, which are the genetic material.

This work was summarized in the September 1957 issue of *The Bacteriological Review*, and earned the Nobel Prize for Tatum, Lederberg, and Beadle. The importance of this work is not to be under-emphasized. It provided a new concept with regard to biological inheritance.

But let Dr. Lederberg tell it in his own words, as he spoke to the Nobel Prize assembly in Stockholm, Sweden in 1958: "That bacteria and their genetics should now be so relevant to general biology is already a fresh cycle in our scientific outlook." And well it has been. Two other Nobel Prize winners, Dr. Francis Crick and Dr. James Watson, were the first to devise a model of the structure of the DNA molecule. They made use of crystallographic pictures from Maurice Wilkins and the late Rosalind Franklin in arriving at the DNA structure. Dr. Linus Pauling of the California Institute of Technology was also very close to making the discovery.

DNA consists of two complementary, threadlike structures wound about each other in the form of a double helix. Along these strands, somewhat like the rungs of a ladder or the steps of a spiral staircase, are placed paired nucleotides, or bases. These are of four kinds and it is the sequence of these entities that determines the hereditary differences between one individual and another. After the structure of the DNA complex was discovered, many researchers in many lands attempted to synthesize this molecule.

Dr. H.G. Khorana, research professor in enzyme studies at the University of Wisconsin, set as his goal the synthesis of a gene. After several years of tireless work, his efforts were rewarded by success. He announced this achievement in the February 16, 1979 issue of *Science* and, like Tatum, Lederberg, and Beadle, became a winner of the Nobel Prize.

Since the middle of the 19th century, it was known that the nucleus of the cell contained chromosomes. As evolutionary biology progressed and organisms became more complex, more chromosomes and genes were necessary. Humans, with very complex bodies and brains, have 46 chromosomes in their cells.

Chromosomes are DNA molecules composed of double helical strands and paired genes. A gene, then, is simpy a segment of the long DNA thread that constitutes the chromosome. A typical gene is a few thousand base-pairs in length, whereas a chromosome may be tens or hundreds of millions. These are the structures which direct and control not only the biological functioning of the cell, but also its reproductive capabilities. How could these chemical entities control and direct both functioning and reproduction? Researchers in many fields have opened the door a little wider, and it has become possible to peer through and to understand with a little more clarity a few of the hidden secrets of life itself.

DNA sits, like Mr. President, in his front office within the nucleus of the cell. He is in complete charge of his factory, the cell. He has at his command an executive officer, Mr. RNA. Should the cell be in need of some specific protein molecules, Mr. President will inform his executive officer of such need. Mr. RNA will carry this message to the factory staff, the ribosomes, where the proteins will be manufactured according to the specifications of Mr. President. The executive officer is given the title "messenger RNA." The translation of the RNA message into a specific protein takes place outside the nucleus and is made with the aid of transfer RNA.

Karyotype of human chromosomes. Scientists use such images to study breaking and relocation of parts of chromosomes.

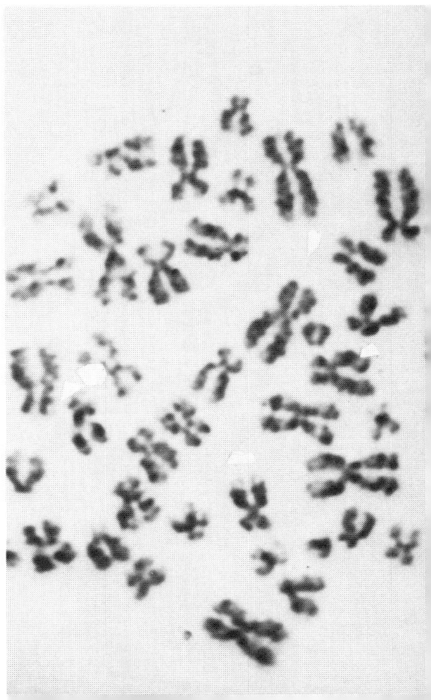

Most living organisms reproduce sexually. The progeny contain a mixture of chromosomes and hence genes, half from each of their two parents. In humans, a sperm has 23 chromosomes and an ovum has 23. When conception occurs these will join to give a total of 46, bringing into the fertilized egg hereditary material from both the male and the female. As the fertilized cell divides, the offspring will have individuality and variations enhanced by having received genetic material from both sources. This is indeed a marvelous mechanism by means of which no two organisms, except for identical multiple births, can ever have the same identities.

Our mythical travelers ask themselves time and again, "Could this development have been accidental? Could these series of sequences have been mere happenstance, or could they have been part of a grand design? If it be the latter, what then is the force, the power, the entity, the source of all that happened throughout these travels?" These questions will continue to be placed before them at each new encounter. As wondrous as these experiences have been, and as beautifully coordinated as they have seemed, only the basics of biological progression have been so far established.

A knowledge of history not only informs us of what has gone before, but can in many instances help us to better understand the probabilities and potentials of the future. Progression in any field of scientific endeavor must be based upon a thorough knowledge of all events which have led up to the current status of that science.With such knowledge at hand, scientists can proceed to develop theories and hypotheses around which future research and experimentation may be planned and related factual knowledge be acquired. Thousands, perhaps millions, of scientists throughout the world have studied the past and have related it to the future. This has been especially true of those men and women who have devoted their lives to the study of the biological sciences. Through their efforts, especially during the last few decades, discoveries have been made which have brought us nearer to an understanding of the very secret of life itself.

We have arrived at a point in biological time where scientific history has provided us with knowledge of the mechanisms so necessary in the great proliferation of living structures and forms and their specific variations. Alterations to the pattern of heredity can be beneficial, as in the production of insulin by bacteria through genetic engineering. Controversial views hold that genetic engineering such as this has the capability of producing monsters. Scientists must settle this problem by carefully controlling research.

We have discussed the mechanisms by which proliferation and variation occur. Ordinarily, however, there is another means by which great variation in a species can be brought about, and that is by the little understood mechanism of mutation. When mutation occurs, genetic material is changed, perhaps by some environmental factor, by abnormal metabolism within the cell, or by a multiplicity of factors which are yet unknown. The offspring are born with features and characteristics different from those of either of their parents. Except for this process of mutation, the greater the number of genes which are contained within specific cells, the more likelihood there is of greater variation.

With all this knowledge at hand, we shall now see what the future has to offer.

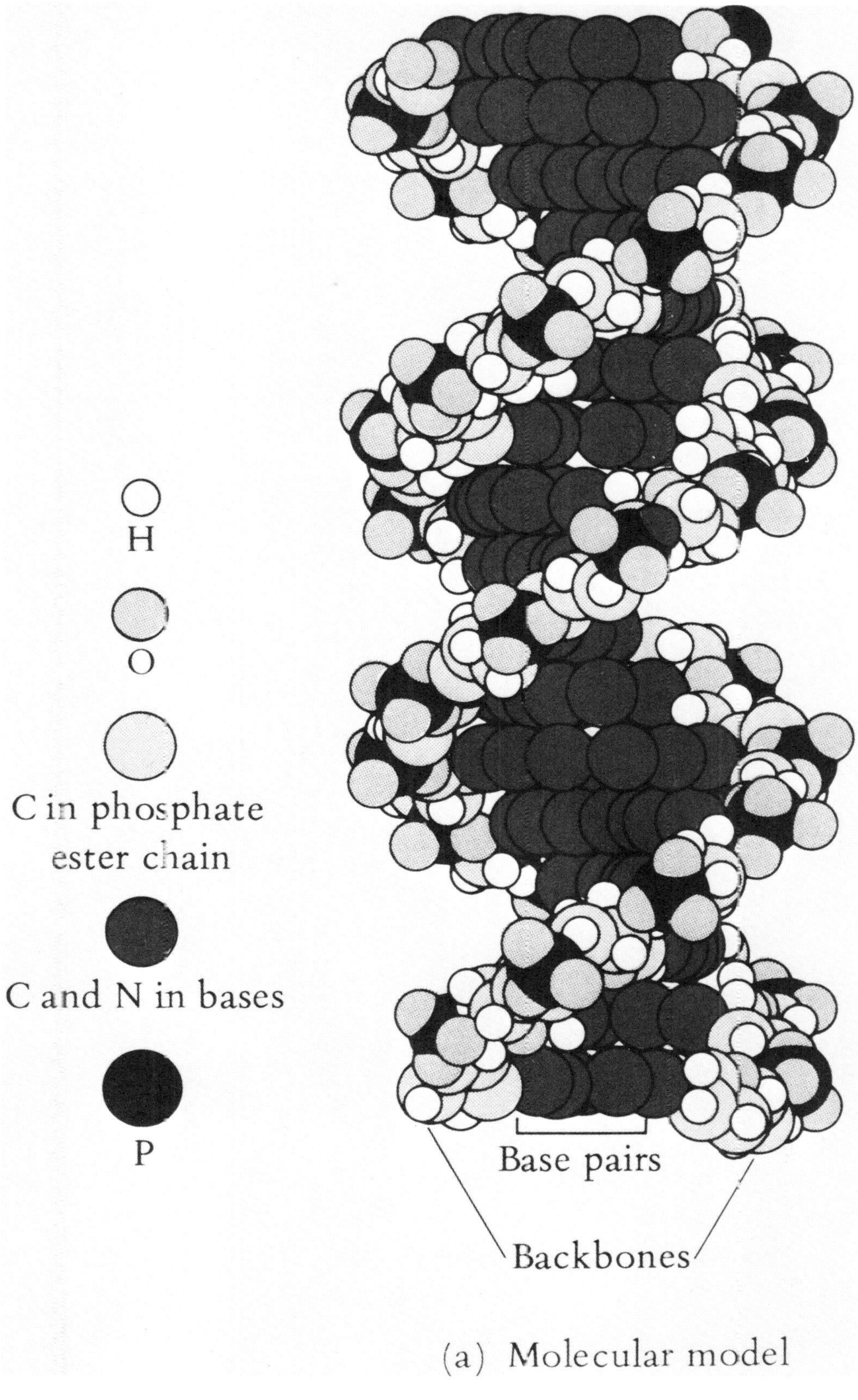

Model of molecular structure of DNA double helix, with a schematic diagram of its components.

Part 3: LIFE PROLIFERATES ON PLANET EARTH

Chapter 13: EVOLUTIONARY CHANGE IS UNDERWAY

All living creatures, from the simplest single-celled organism to the most complex forms of life, share a common need for water, for without water they would cease to function and die. There are a few exceptions to this rule; the fertilized cells of certain plants and animals can go into dormancy during periods of environmental dryness and still preserve capabilities of reproduction until moisture returns. As we noted previously, diatoms have survived a century of such dormancy.

Single-celled creatures, both plant and animal species, reside in marine or fresh water surroundings, or within or on the soil where moisture is available. It is thought that the ancestors of all plant and animal living structures and forms were algae. It can be reasonably conjectured that the evolutionary beginnings of vegetation must have preceded the beginning of animal forms. Algae and bacteria and their plant descendents are the only living things capable, through the process of photosynthesis, of manufacturing food. Animal forms cannot manufacture food, so they survive only by eating vegetation or consuming other animal matter.

It seems likely that, somewhere in the early evolutionary process, certain cells became differentiated so that they lacked this process of photosynthesis, and these became what we know as animal cells. From that point on, two lines of evolutionary development – the plant and the animal kingdoms – were established. However, the same genetic principles concerned with proliferation and variation apply to both kingdoms.

There are many other similarities. Both plants and animals require water, and both receive nourishment from their environments. Both reproduce by sexual means. Sexual reproduction provides not only for the propagation of the species, but also for variation. The union of male and female elements contributes genes from both parents to the offspring. Each offspring is thus an individual with characteristics from both parents, and yet different from either. These variations may be barely noticeable in one generation, but after several generations the changes may be profound.

60 Plants evolved through many variations from underwater vegetation to surface water vegetation and eventually to vegetation that covered the land, providing food, beauty and magnificence, from cabbages and beans to roses and orchids, to heather and giant redwoods. Meanwhile, animals also were evolving. They advanced from single-celled organisms to those which contain several or many cells, to crabs and fishes, to reptiles and birds, and eventually through the mammals to the design plan's greatest achievement, that of humans.

The plant kingdom, important and wondrous as it is, must be delegated to the devices of the botanist, at least for the moment. Intimate knowledge of the intricacies of animal evolution will more completely fulfill the purposes of this book, the most important of which is in some way to bring the minds of humans into harmony with all that is related to their origin and to their being.

We begin then, with the first zooplankton or, as they are oftentimes called, the protozoans. Their many species inhabited watery environments about 1.4 billion years ago. A typical, very primitive protozoan is the amoeba. Amoebas, more properly called "amoebae," have all the makings of a true cell. They can feed on vegetation, bacteria, or other protozoans. They metabolize what they ingest and excrete by-products, and they propagate themselves by sexual reproduction. They are wafted about here and there by currents, but when they are in contact with a solid surface they can accomplish movement by their own very individualistic means, even though they have neither flagella nor legs to assist them. The cell membrane which is in contact with a solid surface bulges out, and the fluid content of the cell flows into this bulge, called a pseudopod or false foot. One such bulge allows movement equal to the length of the cell.

A bacterial cell (Proteus mirabilis) showing numerous flagella.

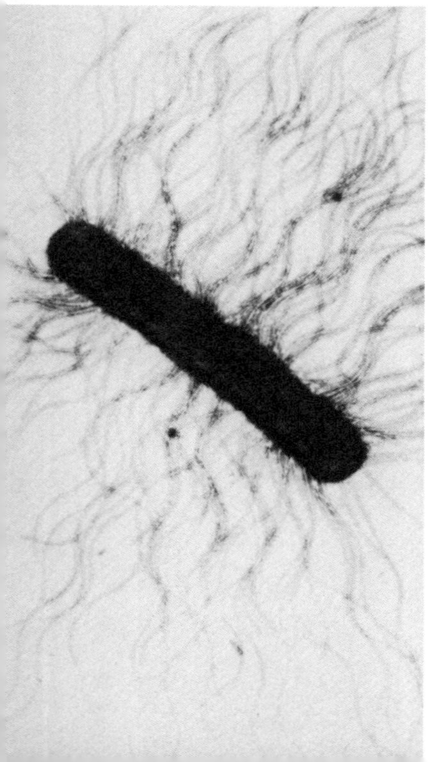

There are myriads of other such single-celled organisms within watery confines. Some amoebas have developed a more refined mechanism for their locomotion, in that attached to their outer membranes they have developed stiff, hair-like projections called flagella. Some biologists are of the opinion that these flagella are modified spirochetes, a primitive organism similar to that which produces the disease syphilis. These hair-like structures are attached to the outer wall of the cell where they have contact with something akin to a muscle fiber. This enables them to move in such a way as to propel the amoeba in a given direction. Other primitive single-celled organisms have on their surfaces soft hair-like structures called cilia. The waving of many of these cilia can also provide propulsion.

Protozoan cells react to light and temperature variations and to other changes in their environments. Although there is no detectable physical structure resembling a nervous system, there seems to be something within the chemistry of the cells that initiates responses to stimuli. These minute cellular structures have within themselves the capability of performing many of the same basic creative functions as the cells of higher animals.

There are 15,000 known species of these protozoans and perhaps many others which, so far, have not been discovered or classified. To observe one of these little single-celled animals under the microscope for a moment is to see it surrounded by a static environment. But what happens when the surroundings change? How these primitive organisms react when multiple environmental factors are introduced!

A good example of this adaptability among simple organisms is the fungus Dictyostelium discoideum, also known as cellular slime mold. One of many small amoeba-like organisms, D. discoideum can reproduce itself by three distinct means. When the food supply (bacteria and other organic materials) is abundant, these fungi remain solitary creatures and reproduce by simple cell division, called binary fission. The two cells which result have identical characteristics and therefore afford no means by which evolutionary variation could occur. Furthermore, if the environment's light, moisture, and ammonia from decaying organic material were not at the proper levels, the fungi would die. But nature has come forth with the most fantastic devices for overcoming the demise of D. discoideum as a species and for overcoming the monotony of self-perpetuation without change in heredity.

When an area which has been occupied by such fungi has been cleared of all available food materials, starvation provides the signal for the fungi to aggregate and to form a multicellular structure known as a slug, the largest of which is about 2 millimeters long and contains thousands of individual fungi. In all probability, this is one of the earliest forms of multicellularity in the animal kingdom. These slugs are capable of migrating for fairly long distances, and when they arrive at a spot where the light is a little more intense and the humidity and ammonia are somewhat less, they progress to the second step in their exotic morphogenesis. They form a fruiting body somewhat akin to a fungus of mushroom, which consists of stalks and a base.

Those cells which lie within the foremost part of the slug go about developing the stalks of the fruiting body, and in the process they ultimately perish. Those cells which are in the rear of the slug have their functions outlined as well, for they have been assigned the important task of ensuring that the race of the slime mold will not become extinct. Certain of these cells remain at the base of the stalk and become known as spores. Here they can remain for long periods of time under drought conditions. When the environment again becomes favorable, these spores come alive and begin the life cycle of the slime mold all over again.

However, these spores reproduce just as their ancestors did, by binary fission, and therefore do not impart to their offspring any new hereditary mate-

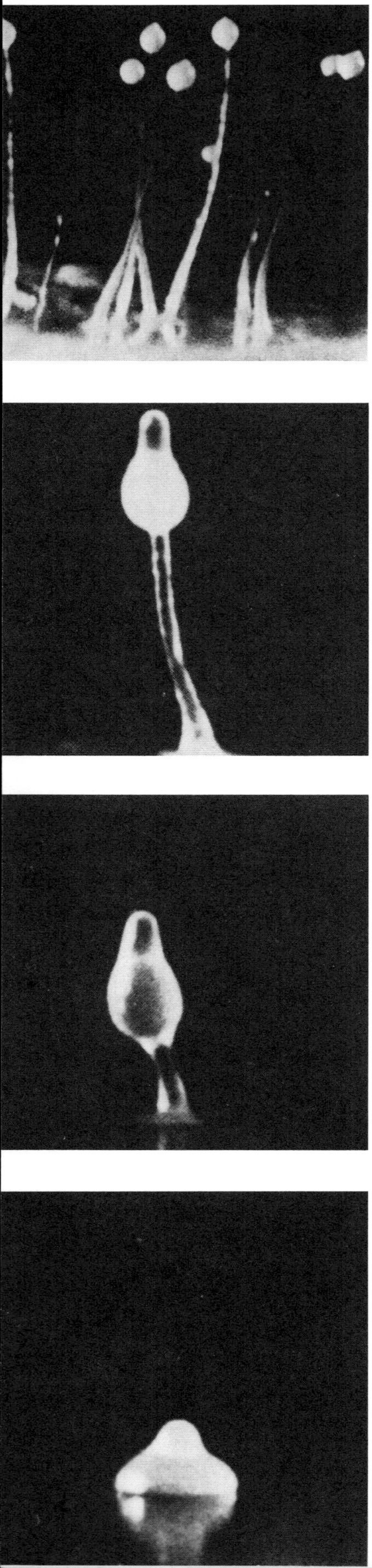

rial. There is still no potential for evolutionary variation. But wait. Do not sell short this tiny bit of protoplasm. Within this early representation of the animal kingdom there is a mechanism whereby certain of the cells in the rear part of the slug become enclosed within a structure, called a macrocyst, which locates itself on the upper portion of the now-dead stalk. The amoebas within this cyst differ from the spores in that they undergo sexual exchange of genetic material. When proper environmental conditions occur, such as very high humidity, these slime-mold amoebas become capable of producing offspring with recombined genetic material originating from both mating partners. These offspring also begin the repetitive cycles of the slime mold.

We have exhibited in this simple form of living substance one of the most marvelous processes, morphogenesis, whereby single cells can aggregate and assume the role of a many-celled organism resembling, in some respects, later evolutionary forms of worms.

Within this slug there also occurred the differentiation of cells working in a cooperative manner to construct this almost vegetable-like fruiting body in which reproduction could occur both with and without the sexual exchange of hereditary material. This series of phenomena is marvelous almost beyond belief, but here it is. It has occurred and does occur. As with so many other things in nature, some of the answers are known, others are not, and some thought to be known may be incorrect.

It is well known, for example, that natural changes of light, temperature, humidity, and the quantity of ammonia present regulate the orderly process through which this morphogenesis proceeds. But what starts it all? What triggers this series of sequences? Due to the brilliant research of scientists such as John Tyler Bonner, Professor of Biology at Princeton University who has worked with this specific organism since 1939, it is known that when all available nutritional material in an area has been fully utilized, the amoebas manufacture a chemical substance known as acracin (cyclic A.M.P.). This substance triggers the D. discoidium amoebas to aggregate and eventually become slugs. Toward the end of the morphogenesis journey, an anti-acracin, a protein enzyme known as acracinase, wipes out the effects of acracin and reset the clocks of amoebic time in such a manner that the life cycle of D. discoidium begins all over again.

Life cycle of a slime mold (Dictyostelium discoideum). At this point a group of amoebas has formed into a slimy mass which begins to pile up in a mound and rise vertically. A stalk is then formed, which carries the fruiting body.

These substances, acracin and acracinase, which were discovered only about ten years ago, have assumed great importance because they are also found in higher animal forms. In man, they are involved as intracellular signals or messengers in the release and regulation of certain hormones within the body, and also with nerve impulses and their conduction.

After witnessing the affairs of this slime mold soil amoeba, our mythical travelers might once again well ask if the profound progression as manifested in this single-celled creature is so simple after all. Their answer must certainly have been "Simple it is not." It has been demonstrated that life in one of its most minute forms has been most complex, mysterious, and miraculous. What has been learned from this microscopic biological structure, the soil amoeba, is that there are chemical and physical interrelationships which have a governing effect upon structure, form, and function, and that nature has devised a clever series of sequences to permit these characteristics to become manifest. Not only does the amoeba benefit by such built-in mechanisms, but all future biological entities do so as well.

Part 3: LIFE PROLIFERATES ON PLANET EARTH

Chapter 14: BIOLOGICAL EVOLUTION CONTINUES

Nothing in the universe is static. Planet Earth inherited the principle of continuing change and imparted it to the inhabitants of Earth. So far these writings have concerned themselves with those facets of biological evolution which have made a positive impact on what future living structure, form, and function would be. However, along with this forward progression from the chemical inanimate to the animate, there must have been some evolutionary failures. There must have been some trial and error until proper combinations were found. For example, out of the tremendous number of possible chemical unions which could have resulted from 92 elemental atomic entities, only a few were destined to contribute to the creation of living matter through a chain of sequences of methane, amino acids, protein, and nucleic acid molecules.

Even though, in this our 20th century, at least 15,000 species of protozoans are known to exist (and many of them have been perpetuated from the earliest days of their species' existence), it can be readily conjectured that many more species might have perished. Some species might not have been geared to perfection for the continuation of life's processes. Some might not have been capable of adapting to changing environmental surroundings. Some which were weak might have been eased out of their favorite spots or even gobbled up by their stronger neighbors. Those which have survived were able to do so because of their genes. Their inheritance made it possible for them to adapt and thrive; this is what biological evolution is all about. To put it in Darwin's words, it is the survival of the fittest.

The process of simple cell division was later replaced by that of two mating cells coming together and exchanging some of their genetic materials in order that the offspring might take some hereditary qualities from both mating partners. This process was not destined to accomplish total exchange and therefore the variations from generation to generation were somewhat limited. These methods were too slow to meet the design plan's biological time schedule. If, within a billion years, the evolutionary process was to proceed from amoeba to humans, a much better method of reproduction must be devised.

66 There came upon the scene an organism called *Alga volvox*, in which certain cells were set aside to accomplish perpetuation of species. The method of reproduction of *A. volvox* differs greatly from the simple mating-partner method of the soil amoeba. In the volvox, certain cells became positive and negative, male and female, or, in other words, egg and sperm. One cell, the ovum or egg, becomes rounded and remains stationary. The other cell becomes elongated, acquires motion, and assumes all of the roles of the sperm cell. It can seek out and contact the egg, penetrate it, and carry to within its nucleus the inheritance potential. Following the male penetration of the egg, the chromosomes of both unite. When the first cell division of the fertilized egg occurs, a complete gene assembly is present.

By this process the maximum potential for inherited characteristics was achieved, and the creation of more complex living structures and form proceeded more rapidly. Once again evolution provided perfection, a positive advance toward obtaining the ultimate goal of creation. Along the pathway leading to this goal all future created living things, both plant and animal, will propagate by the sexual union of male and female components. As Marilyn Monroe once put it, "Sex is a good thing. I think it is here to stay."

So far, emphasis has been placed upon the evolutionary progression of those minute single- or many-celled microscopic organisms which lived pretty much by themselves as individual species. They nourished and maintained themselves by what they could gather from their environments, preying upon the weak, with the strong surviving, more adaptive to their surrounding space.

But there are other types of minute organisms which prey upon others in a more selective fashion. These are the parasites. Oftentimes living in complete harmony with their hosts, frequently they are capable of supplying one or more components necessary to the host's continued existence, while at the same time taking from the host those substances which are essential for their own well being. Some parasites, however, are indeed deleterious, and can completely destroy their hosts.

It is a common belief that bacteria, viruses, amoebas, and parasites are bad actors, and that the less we have to do with them the better off we are. As a matter of fact, only a few of these organisms are capable of producing disease. Since many of these organisms are capable of reproducing every twenty minutes, starting with only a single pair, in just three hours there would be more than a thousand. Carry this rate of reproduction through days, weeks, months, and years, and you can readily understand the massiveness of any single species. Even though as individual units they are microscopic, they would become so great that the planet would be overwhelmed by any one group. No other living thing could survive.

However, these organisms prey upon each other and keep their populations under control. There are other benefits. Certain of these single-celled organisms fix nitrogen in the soil, a process which enables plants to utilize it. Others cause wine to ferment and bread dough to rise. And without certain bacteria in the intestinal tract of higher animals, including man, certain enzymes necessary to the digestive process could not be manufactured.

There are many instances of symbiotic tenure in which one or more organisms work in unison to promote the well-being of the whole. When the true cell came into existence the cell became inhabited by an algae-like structure, and bacteria modified themselves to become the organelles called mitochondria. Parasitic symbiosis, where entire organisms inhabit other organisms for the common good, supports the idea that perhaps nothing in life can be completely independent. All biological forms, from the first true cell to the highest complexity of living matter, share this interdependence.

Coral has, within its polyp, algae cells which use the carbon dioxide and water waste material from the polyp to manufacture, by the unique process of photosynthesis, starches and sugars for its own well-being as well as that of the polyp. There are many other instances of such essential togetherness. One very interesting relationship is described by Lewis Thomas in his book *The Lives of a Cell*. He describes the important role which a single-celled protozoan called *Mixotrica paradoxa,* a true cell composed of symbiotic components – plays in the well-being of a higher organism. It is a parasite found in the termite ant's intestinal tract, where it performs a function which the ant is incapable of performing itself. Termites feed on wood, but they do not have enzymes within their systems to enable them to digest and break down cellulose. *Mixotrica paradoxa* has the ability to swim about and manufacture the enzyme which reduces cellulose to starches and sugars to be used by both mixotrica and the host ant. And, in addition, a by-product known as lignum results, which the termite ant can excrete in the form of pellets which it in turn uses to construct the wall and the arches forming the intricate compartments of its nest. Nature has been marvelous indeed in providing such a social arrangement by means of which both the parasite and the host benefit. The same natural forces which created and guided all living things from simple bacteria to complex man also established that interdependence, both physiological and social, is essential for life's continuation.

Boris Pasternak wrote in *Dr. Zhivago*, "If you want to know, life is the principle of self-renewal, of simply renewing, remaking and changing, and transfiguring itself." The greatest changes in evolutionary progress have come about since the introduction of male and female sexual components. From that point on no two individuals, whether protozoans or humans, would ever, except for identical multiple births coming from a single egg, be identical. Is is genes that have made it all possible. If it had not been for the initial evolutionary development of deoxyribonucleic acid, the double-helix strands along which are contained the paired genes, all of that which our travelers have witnessed so far, from bacteria to algae to amoeba to *Mixotrica paradoxa*, would never have come to be, and Earth would have consisted of nothing other than itself.

Genes account for the biological past and the biological present. Each generation will continue to contribute its unique changes to that of all future evolutionary development. As time presses forward, this ever-changing universe will continue to express itself on Earth in multitudinous ways. In the biology alone, changes will be almost unbelievable.

The last billion years of Earth's existence confront our group of travelers. Now, we shall advance a few hundred million years. Our mythical travelers and those of us who follow will be once more assured that things which could not be even dreamed have a way of becoming real.

Part 3: LIFE PROLIFERATES ON PLANET EARTH

Chapter 15: SEARCHING, WINNOWING, AND SIFTING

Even a blind dove can find a pea now and then. So goes one of the clever sayings of your author's late, beloved mother-in-law. So it goes with those beloved scientists, the archaeologists, the anthropologists, and the paleobiologists. Not that they attack their problems blindly, but that they encounter so many frustrations along the way.

For example, relatively few species of all the vast number which have evolved within the animal kingdom have survived to the present. Richard C. Lewontin, in an article entitled "Adaptation," in the September 1978 issue of *Scientific American*, states that there are about 2 million known living species. It is estimated that 99.9 percent of all species that have evolved since the first chemosynthetic bacteria are extinct. All species had the potential for survival, however, only one tenth of a percent of a possible 2 billion have, in fact, been able to survive. The others lacked the requisite genes to enable future generations of their kind to cope with changing environmental demands. Change has ever been the pattern not only of biological evolution, but of the physical world as well.

In 1859 Charles Darwin, in *The Origin of Species,* defined the principles upon which natural selection could occur. Within a given species there were variations in structure and form. These variations were capable of being inherited. Lastly, there was a struggle for survival in which the organism that could adapt to changes in its environment would survive while those that could not would perish.

Natural selection and the survival of the fittest are somewhat easy to accept. However, we are incapable of explaining all of the intricacies of adaptation whereby structures are specifically and beneficially modified to solve environmental problems.

The theory appears to be sound. Where there are organisms that can thrive in a given environment due to structural modifications, and that have closely related species without such structural modifications that cannot thrive, is the modified organism a descendent of the one that could not adapt? Are those organisms with several structural modifications the result of cross-breeding between somewhat unlike members of a species? If the answer to each question is yes, then how did this information come from the genetic code, so that all future generations would benefit by those particular structural modifications? Darwin could not answer questions regarding the inheritance of acquired characteristics. However, he possessed exceptional capabilities. He was able not only to observe, but to interrelate all that he saw among living creatures. He witnessed closely related members of a species that had acquired structural modification and behavioral patterns thriving in an environment that was somewhat different from that in which their ancestors had lived.

Throughout all of evolutionary history, there is evidence that structure and form were modified in a manner that enabled organisms to function according to their needs. For example, a marine amoeba, wafted from place to place, has a changing environment and thus its food sources are not as easily depleted as they would be in a stationary existence. Yet, when this amoeba became earthbound in the soil, it had to provide its own means of locomotion. It developed the pseudopod method, indeed a primitive means of locomotion, but sufficient.

More refined types of locomotion came when the amoebas developed flagella or cilia which wave about and propel the organism. During the last billion years of earth's existence, when structures became multi-cellular, extensions from the outer layer of cells became flippers, arms, or leg-like structures which enabled the organism to move about on its own volition.

Such modifications in an organism can be readily extended to the fins of fishes, to the legs of dinosaurs, to the legs and wings of birds, to the four feet of mammals, and lastly to the complex functioning structures of arms and legs of humans. Of course, there are many more structural changes not related to motion.

Evolution has progressed from the simplest, single-celled organisms to humans, in which billions of cells of some two hundred varieties have gathered together to perform a number of most intricate functions. The number of structural modifications required to meet these needs seems almost beyond our comprehension.

But we can be assured that they have occurred, otherwise there would have been failure in our abilities to adapt. One of the most profound examples of this evolutionary history is demonstrated by the embryonic history of a human infant.

With conception, there is a union of the male sperm with the female egg. Within the nucleus of the egg a joining of chromosomes occurs. There were 23 chromosomes in the egg and 23 chromosomes in the sperm, but when the first cell division takes place each offspring cell has 46 chromosomes, which carry with them the inheritable qualities of both parents. In addition, individual cells must get their nourishment from their environment. There is, in all probability, some signal which causes them to continue to multiply and to aggregate. This process goes on while the cell is being propelled, not by its own flagella, but by the cilia on the cells of their passageway. By the time they enter the womb, the aggregated cells are 16 in number and still capable of receiving adequate nourishment by simple absorption. However, as the number of cells increases geometrically, there comes a time when simple absorption does not answer the needs of this cell grouping. Evolution again repeats itself, because this globular mass of cells surrounding a cavity becomes parasitic and attaches itself to the wall of the womb. There it can absorb from its host, the mother, those substances necessary for its continued viability. Yet more growth takes place and there comes a time when this simple attachment is again insufficient, and then another unique structural modification comes into being. It is not known just exactly how this occurs, but certain cells within this same globular mass become differentiated to form the placenta, capable of nutrition and waste exchange between the developing embryo and its mother. This is an advance in function over egg-laying creatures, where the nutritional requirements of the developing embryo and fetus are contained within the egg itself in the form of the yolk. Now, the entire process of creating a living form which can exist in the external world occurs within the maternal structure, the uterus, which has evolved for this purpose alone.

Fluid is contained inside this bundle of cells. Here the embryo develops in a marine world all its own, but with still further modifications in structure which enable it to cope with its special surroundings. Cells differentiate and multiply to form organs, but while doing so, the embryo remains attached to the placenta by means of the umbilical cord. This cord contains arteries and veins to ensure contact with its host for nutrition and waste disposal. This is not so different from that which occurred in its evolutionary ancestral past.

Human embryo of about seventy days. Notice the enormous development of the brain at this early stage.

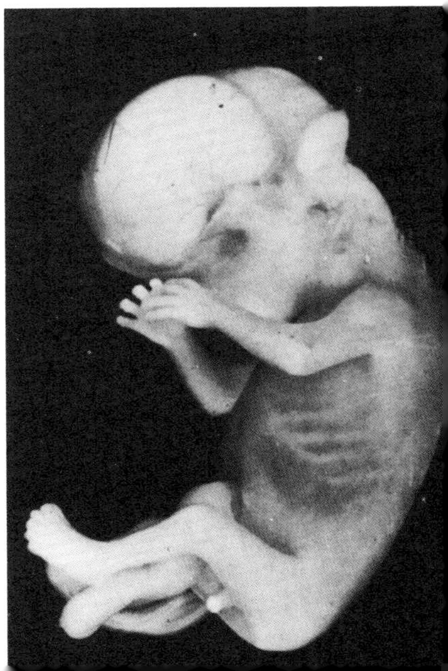

As the embryo develops, certain cells become differentiated to form three clefts or gill slits on each side of the foremost portion of the embryo, where the head will eventually develop. These gill slits, useless as they are, form part of the evolutionary heritage from gills, a structural modification of the fishes. From these gill clefts will subsequently develop the nose, mouth, dentition, wind pipe, and lungs, for there will come a time when these structural modifications will be necessary. When the infant is born, it must adapt to an environment containing not a surrounding fluid, but air. It must be able to breathe and absorb oxygen from the air.

Other early embryonic modifications also mimic those of our evolutionary past. The limb buds, the precursors of the arms and legs, are not too dis-

72 similar in both placement and function to the fins of fishes and the flippers of turtles and dolphins. Just as surely as these initial outjuttings of flesh have developed in other creatures into highly specialized structures, such as the arms of the crab or the lobster, so will these limb buds in the human embryo become arms and legs capable of the most sophisticated functioning. Does this all seem silly, to refer to a repetition of the evolutionary past in the embryonic phases of the development of a human being? Many people think so. But there is no denying that it is there, becoming manifest in the early intrauterine life of each new human being. Biological modifications have occurred, both purposely and meaningfully, to meet the changing requirements of individual organisms, to respond to changing environments, and to meet the specialized needs of more complex organisms.

These modifications are passed on through the genes to future generations. But, let us be clear that structural modifications to meet functional requirements do not occur overnight, nor from one generation to the next. However short or long were the periods between species differentiations, let us remind ourselves of the great span of geological time. It took some 1.2 billion years for the Earth to prepare itself for the advent of life – for just the first chemosynthetic bacteria – and another 2 billion years before the first true cell evolved. It was yet another 1.4 billion years before the first toolmaking precursors of humans came upon the scene. Humans are newcomers to planet Earth, and the time during which they have occupied it is comparable, on a time chart, to the space taken by a floating seed compared to the whole sea.

Now, we return with our mythical travelers to that place in time about 850 million years ago, when significant numbers of structural modifications affected the course of evolution.

Part 3: LIFE PROLIFERATES ON PLANET EARTH

Chapter 16: FOSSILS AND THEIR MEANING

It was the year 79 A.D. Vesuvius shook, split asunder, and roared once again. A vast vapor cloud ascended into the sky, followed by a continuous series of explosions thrusting upward masses of ash, molten lava, and huge flaming chunks of Earth's innards. Within a short time the skies darkened and two cities, Pompeii and Herculaneum, with their inhabitants and all of their possessions, were engulfed in wave after wave of the super-heated products of volcanic action.

Both cities were completely covered during this catastrophe, Herculaneum to a depth of 65 feet, Pompeii to a somewhat lesser depth. At first there were opportunistic and destructive excavations of these cities. Later, more scientifically planned and executed searches resulted in much of these two cities standing today almost as they were then. Mount Vesuvius, by its eruptive action, preserved for scientists to study, and for worldwide visitors to see, one portion of our historic past.

Preservation, then, has been the key to our understanding of the past, and of the time relationships of the non-biological and the biological manifestations within and upon Earth.

In these writings, our chief concern is with the earliest of the begetters and the begotten. These biological entities have changed by increasing complexity to eventually become the only creatures with functional structures which enable them to study in detail, to reconstruct evolutionary history, and to look backward and relate in time all that has gone before. Humans alone are able to say, "These are the things of which I am made. These are the historical and factual circumstances without which I could not have been. And the great design plan has been followed to the letter in bringing about my creation."

But how do humans know all of this? How can we be sure that these ideas are something more than some idle philosophy? Just as Pompeii and Herculaneum have been preserved, there are fossil remains of the animal kingdom, from the simplest to the most complex. When put together in order they reconstruct this story in a most conclusive way.

What, then, are fossils? Fossils are either the physical remains of once-living organisms or physical manifestations that these organisms once existed in given surroundings. They may be found wherever the organism was present. They have been preserved by having been surrounded by various types of precipitate, usually forms of calcium or silicate, which in many instances has hardened into stone. Other organisms have been preserved by the infiltration of mineral substances, and in this way they have become impervious to the destructive forces of their surroundings.

Upper) - During the Mesozoic age, ammonoids became the dominant animals of the seas.
Lower) - The most numerous of all arthropoids during the Cambian age were the trilobites. They molted often, and their many discarded exoskeletons were fossilized. Two longitudinal furrows divided their bodies into three parts, which gave them their name.

The bones of fishes and the outer structures of shellfish have recently been found embedded in sediment which eventually became stone. Where the actual organism has not been preserved, there are sometimes cavities or molds of the organism, much like a footprint, where the perishable material itself has been destroyed. For study and preservation, casts of such molds can be made with either plaster of Paris or latex. The details of the structure can then be studied.

On Sanibel Island, off the west coast of Florida, fossils of shellfish are constantly being washed ashore. By studying ocean floor tracks of the burrows of worms and other organisms, fairly good knowledge of the nature of these biological forms can be obtained.

What is now a fossil was once a living creature. When its remains or traces became embedded in sediment or rock can be accurately determined by radiometric means. The relative date can be set by timing the degradation of uranium into radioactive lead, of thorium into strontium, or of potassium into argon and, for dates within the last 50,000 years of Earth's existence, by carbon-14 technology.

In addition to setting the time at which such structures existed, we can observe the comparative anatomy of these organisms by studies of shells, bones, and other structures and forms. Comparative anatomy, when related to the age, reveals a progressive evolutionary pattern.

The farther back we go in evolutionary history, the less fossil evidence we have. As evolution advanced to the shell and to bony-structured individuals, more fossils are available. However, there is evidence from the fossil remains found in the coral reefs off the coast of Southern Africa indicating that the simple methane-producing bacteria, which have been described in previous chapters, existed as far back as when Earth was 1.2 billion years old. In western Australia, in rocks 2.7 billion years of age, there are embedded two distinct types of fossils of biological materials. One resembles the spicule of a sponge, and the other a fragment of a starfish. In the latter, there is a series of microscopic globular masses which appeared to have been undergoing cell division.

In 1954, the late Stanley Tyler of the department of geology at the University of Wisconsin-Madison, while searching the Gun Flint area off the northern shore of Lake Superior, found strange rock formations. They appeared to be layer upon layer of sedimentary material which had hardened into what are known as stromatolites. They were dated radiochemically as 1.4 billion years of age. These stromatolites were found to be of non-marine origin. Tyler's studies of these rocks revealed fossil remains of bacteria, of the true or eukaryotic cells. He also found a wormlike structure which contained a central tube running from one end to the other, forming a digestive tract. These findings were, to date, the oldest fossil remains of the true cell and of a wormlike struc-

Fossilized marine animals including clam shells, found in Wyoming in a Jurassic period rock.

ture. Tyler's findings not only indicated that these cells and this wormlike structure existed 1.4 billion years ago, but they had been land-dwellers rather than sea-dwellers.

There is a large gap in the fossil record between the time represented by Tyler's fossils and the appearance of large numbers of multicellular, soft-bodied creatures. Two important geological discoveries, however, have helped greatly to clarify this portion of evolutionary history. In the July 1979 issue of *Scientific American*, Simon Conway Morris, in an article titled, "The Animals of the Burgess Shale," describes a find of fossil remains which has been most important in helping to clarify this previously hidden time of fossil record. This time lies between that of the creation of the many species of the single-celled living structures known as the protozoans and the beginning of that period when multicelled creatures developed hard parts, such as external shells or internal skeletons.

76 The Burgess Shale area of western Canada has provided numerous fossil forms of many species, some not previously classified, and multicellular forms prior to 700 million years ago. The Burgess Shale deposits have been dated at about 530 million years of age. Morris mentions an area in southern Australia where fossils of some species which have been found in the Burgess Shale area have also been discovered. These shale deposits were first discovered in 1909 by Charles Doolittle Walcott, then secretary of the Smithsonian Institution. Huge pieces of rock had fallen from a vertical incline 75 feet above, and embedded fossils were found in both the fallen rock and that which remained intact.

On two visits, Walcott sent back to the Smithsonian Institution over 1,000 pieces of such rock. Studies of these fossils have continued from that time to the present, and over 120 species of multicellular, soft bodied creatures have been classified. Among the fossils discovered were specimens just one category ahead of single-celled protozoans, and the gap in knowledge of evolutionary progression was closed.

The southern Australian fossil-bearing deposits were some 70 million years earlier than the Burgess Shale deposits, and they did not contain some of the earliest specimens of soft-bodied creatures that were discovered in the Burgess Shale. In the latter, there were found fossils of eight known phyla, some of which are present among today's living creatures. About ten phyla, now unknown, evidently perished along the pathways of evolutionary progression.

Among the specimens, aside from the sponges, there were the first structures which had adapted by forming a digestive apparatus, either a pouch or a tube. Among these were jelly fishes, sea urchins, and corals. Next came the echinoderms, which include the starfishes, the sea cucumber, and the crinoids or sea lilies. Lowell R. Laudon, Emeritus Professor of Geology at the

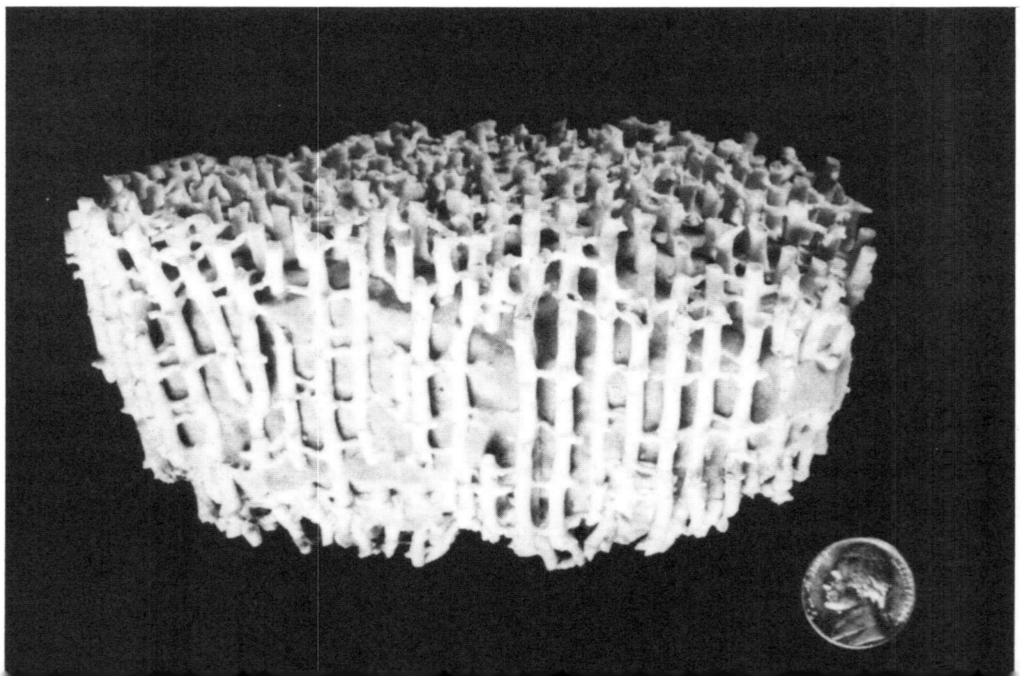

Fossilized sea coral. What we think of as coral is the hard outer skeleton of a colony of creatures called anthozoans. Each animal, of polyp, consists of a hollow container made up of layers of skin and jelly. When alive, each polyp secrets its own calcium-containing skeleton. Large groups of these skeletons make up coral reefs.

University of Wisconsin-Madison, whose fossil collection is widely recognized, states that these sea lilies or crinoids grew to be as big as a large-sized living room. The specimens included oysters, jellyfish, and also arthropods, the first creatures to develop jointed extremities. There were also specimens of segmented and non-segmented worms, of the three-lobed creature, now extinct, known as the trilobite, and also of brachiopods. (Similar fossil specimens can be found on the shores of Lake Michigan in the Door County Peninsula of Wisconsin.) These soft-bodied creatures later developed into the squid, the octopus, and the crab and lobster.

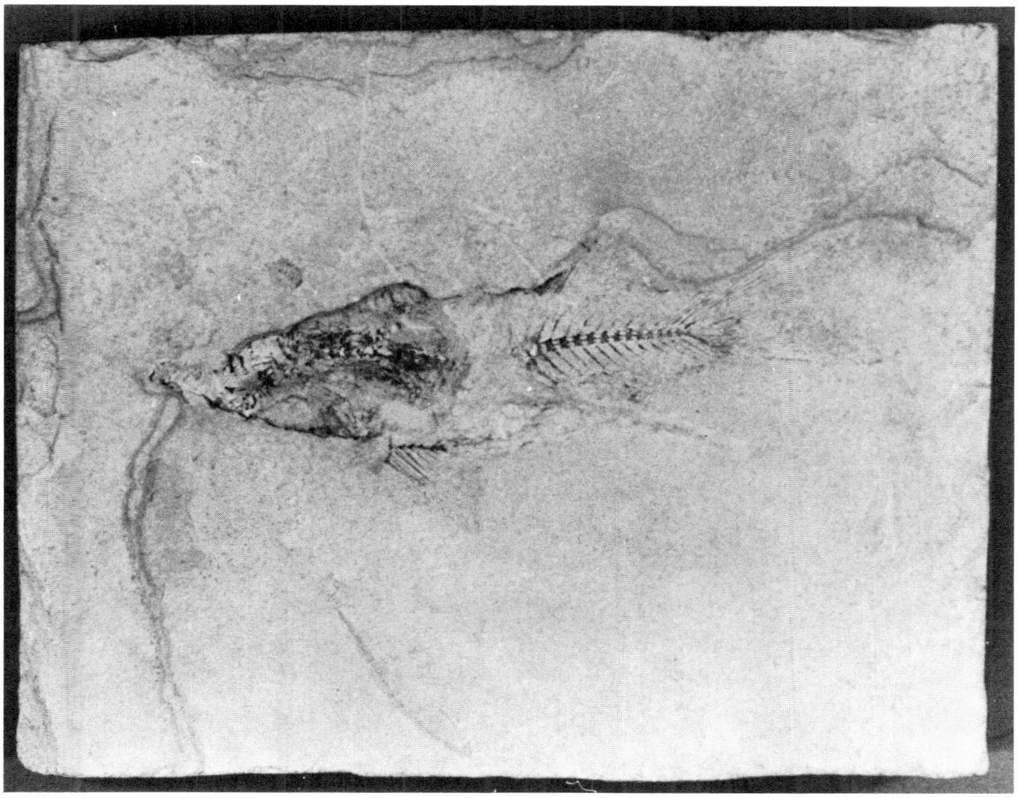

Fossil of a nearly complete fish skeleton, found in the Burgess Shale. The shale has been radiometrically dated at 537 million years.

In the Burgess Shale, there was also found the earliest known invertebrate which immediately preceded the vertebrates, animals with a bony spine. No fossil-bearing geological deposit ever discovered has contained relics of the past so necessary to the understanding of a period of biological evolution. Here, similar to Vesuvius, some catastrophe had enclosed, within the precursors of a rock formation, the fossil remains of creatures which lived half a billion years ago. This event has preserved for us the story of the evolution of living structure, form, function, and adaptation, from the single-celled protozoan to the creation, advance, and modification of multicellular, soft-bodied creatures living within a marine environment. Furthermore, all of this information, enclosed in a casket of stone, was thrust up from the bottom of the sea to high above its surface. There it became available to humans, who studied it and made a positive contribution toward understanding. It informed thinking men and women about one phase of their ancestral past, and one portion of the plan which was designed to bring them into being.

From now on, because of the Burgess Shale, evolutionary history, though it grows somewhat more complex as it advances, will become in many ways easier to understand.

Part 3: LIFE PROLIFERATES ON PLANET EARTH

Chapter 17: FROM SPONGES TO SPIDERS

Sponges to spiders. What a marvelous universe this is in which we live. There is time, there is space, there is motion, and there is substance or mass. According to Albert Einstein, these are all relative, and their combined interaction produces energy, some intelligible forms of which are heat, light, radioactivity, and gravity. All of these have become manifest in the universe as we know it, whether it be the union or the splitting of atoms, or the mathematically precise control of the orbits of the components of the solar system.

Planet Earth is a product of elements and forces. Among the many billions which exist in other solar systems of other galaxies, it is the only planet we know of which has been designed for the purpose not only of creating life, but of making it possible for such life to change, evolve, and adapt.

Our mythical travelers, both encumbered and enlightened by all that they have encountered, are now ready to go forward into the last 800 million years of evolutionary history. For, of all the time since the first living matter came into being 3.4 billion years ago, this last one sixth of the total chronology of Earth's existence provides a series of biological progressions of which the most imaginative minds have never dreamed.

From sponges to spiders! Now we shall encounter a few of the creatures which lived during the period from which we have fossil remains. We will learn how versatile they became, how well they adapted, and what roles they played in a progressive evolutionary pattern.

Stanley N. Cohen, a molecular biologist and Chairman of the Department of Genetics at Stanford University Medical School, and Dr. James A. Shapiro, Professor of Microbiology at the University of Chicago, in an article titled, "Transposable Genetic Elements," in the February 1980 issue of *Scientific American*, describe the results of research which may have an important bearing on our understanding of the laws of inheritance. They demonstrated that the rules governing direct chromosomal genetic recombination whereby offspring inherit characteristics from both parents can be bypassed by joining together segments of DNA which are unrelated to the genetic composition of either parent. This research into the realm of genetic engineering is a step forward in our knowledge of how variations and adaptive structural changes can occur in living organisms.

Some of these inheritable characteristics and mutations have been favorable to the species and others have not. Those creatures which benefited have survived and have become a continuing part of biological evolutionary history, while those which inherited unfavorable or disadvantageous characteristics could not cope with their environments and perished.

To demonstrate the point that only those organisms which benefited by their inheritances could survive as the fittest, let us examine some of these multicellular, soft-bodied creatures of 800 million years ago and determine in what manner these inherited structural change were favorable to them.

It was necessary that two basic needs be met; the first was to obtain the nutritional requirements for life – to be able to ingest food and discharge waste products. The other was for protection against environment and predators.

We shall begin with the first of the multicelled structures, the sponge. Adaptations appeared that were beneficial to the sponge: in order for the sponge to become stationary in its marine environment, certain cells became differentiated, forming an attachment to rocks or other solid substances on the ocean floor. But this caused new problems. If there were no rigidity, this cellular mass could easily be disintegrated by the motion of the water. In the case of the sponge, this problem was solved by the absorption of calcium carbonate and silicates, which became spicules of support. Since the sponge cannot move about to obtain nourishment, nourishment must be brought to it. The cells of a sponge are arranged in the form of tubules which draw in nutrient-containing water. Then, after the sponge has absorbed the needed nutrients, water is expelled at the opposite end. Thus the lowly sponge can stand upright, protect itself from its predators because of the hardened spicules, and provide for its nourishment in its own fashion. There are 3,000 species of sponges varying in detail but fundamentally the same.

In order, the next adaptive structural arrangements were food gathering, food ingestion, and digestion (in contrast to the earlier mechanism of absorption of food). These developments were made by a group known as the coelenterates. The name means "hollow intestine." The jellyfish is a good example of this group. It contains a pouch surrounded by a mass of jelly from which trail several string-like structures. This organism can float about on a watery surface, touch and sting its prey, and then literally sit upon it, ingest it, and digest it. This adaptation is not totally new. Several members of the single-celled protozoans had minute organs which they utilized for this purpose. However, the mechanism by which the jelly fish can float about and seek its prey is a new adaptation among the soft-bodied creatures.

Each successive species has preserved something from that which has gone before. For example, the starfish, the sea cucumber, and the sea lilies, a group with spiny skins known collectively as the echinoderms, have borrowed from the sponge their thornlike structures of calcium and silicates which protrude from the skin and provide protection. The most mobile of this group is the starfish, which is not a fish at all, but has unusual adaptive qualities. It can move about freely, it can sting, paralyze its prey, and then, after the fashion of the jelly fish, can ingest the prey in its pouch. One amazing inherited, adaptive quality is the starfish's ability to sit upon an oyster shell and, by means of little suction cups on the underside of each radiating arm, pull open the shell of this bivalve against the strong action of the oyster's muscles which keep it closed. After doing so, the starfish engulfs the oyster in its digestive pouch and leaves behind the empty shell.

In these first three categories of 800 million year-old creatures, we have seen adaptive structures for gathering and digesting food, for protection, and for motion. Now, we shall look at some of the more advanced species of these soft-bodied organisms.

There are several thousand species of flat and round worms. These are elongated structures consisting of from a few to many segments. They have primitive eyes which are very sensitive to light. They use a squiggly type of motion. All are equipped with a tubelike digestive apparatus. Several species, such as the parasitic tapeworm of man, have sucker-like mouths so they can attach themselves to an organism and extract nourishment. The common earthworm is one of the most advanced of the worm species in that it has a circulatory system and both male and female sex organs. The male organ is enclosed by the front of the worm and the female organ is enclosed by segments further back. This poses a unique problem for fertilization. Earthworms have adequately adapted to this, for when they emerge from their burrows, especially at night, two earthworms will lie alongside each other, but with their heads at opposite ends, so that the male organ of one earthworm will be in just the right position for the female organ of the other. Eggs are extruded in saclike formations, where they remain until they emerge as worms.

Another adaptation to its own needs and environment is that which is manifested by a species of mollusks. Mollusks, you will recall, are the shellfish, including snails, clams, oysters, squid, and octopus. Most of these soft-bodied creatures have enclosed themselves within some sort of protective shell. Snails and clams can extrude a muscle portion of themselves, called a foot, and literally drag their homes with them. Oysters and barnacles fix themselves to rocks or other permanent sites shortly after emerging from the egg, and there they remain for the duration of their lives. Occasionally they will attach themselves to the shells of other organisms and get a free ride.

Squids and cuttlefish have a special mechanism for accomplishing motion. They are capable of taking in water and suddenly ejecting it in jet-like fashion, trailing their multiple tentacles behind them. This is sometimes accompanied by the expulsion of an ink-like substance which clouds and darkens the water. It probably is a protective mechanism, for then potential predators cannot see them.

Of all the soft-bodied creatures, crustaceans have evolved to the greatest degree. This category includes shrimp, crayfish, crabs, and lobsters. They have developed an external covering known as an exoskeleton. This is not really hard; it is composed of a somewhat flexible substance called chitin. The chitin over the vital organs of the head and the thorax of the chest is somewhat thicker and firmer than that over the other portions of the body. The body of these animals consists of a distinct head, a thorax, and a segmented tail, although, in some cases, the head and thorax are fused into a cephalothorax. All of these creatures have complete digestive systems with mouth, stomach, intestine, and anus. Each individual has a pair of eyes, one on each side, for seeing on each side of the body. There are five pairs of jointed legs, with one pair for walking, and one pair modified for grasping. The flipping action of the last segment of the tail, especially well developed in shrimp, crayfish and lobsters, produces rapid backward-swimming motion.

A fiddler crab venturing forth from its burrow.

Let us consider the great step forward which has been made in the food gathering and digestion adaptations. On each side of the foremost part of the body there are filament-like extensions which are highly developed organs for the detection of food, prey, and contact. And there are two highly developed, jointed, arm-like structures: one on each side. Each contains a marvelously designed claw for picking up food, for crushing it when necessary, and for transporting it to the mouth. Crustaceans have complete sex structures for the propagation of their kind. They are complete, self-reliant creatures, most complex in structure, design, and adaptation. And, already at this period of biological evolution, they are equipped to perform most of the functions which later higher organisms will find necessary.

Certain crabs have utilized adaptive structures in the most fantastic manner. The fiddler crab is capable of moving rapidly in any direction and retreating to its burrow, waving its arm-like claws in a menacing manner against any who might intend to invade its territory. The hermit crab occupies the shell of a defunct snail, changing its shell habitat as frequently as growth demands.

Most interesting of all is the unique performance of the decorator crab. Mary K. Winston, research marine biologist at the University of Southern California, has studied, both in nature and in the laboratory, the habits of several members of this family of crabs. The decorator crab is so called because it selects and decorates itself with all sorts of material for the purpose of camouflage. The tentacles can feel, the eyes can see, and the brain can direct the jointed arm with its claw to choose and transfer to the mouth all kinds of odd colors, shapes, and forms from its environment. Objects placed in the mouth of the decorator crab are roughened on both the surface and edges by the crab's teeth. When an object has been prepared in this manner, the crab retrieves the piece of material and places it upon its back or on the upper portion of its legs, where microscopic, hook-like structures receive it and fix it in place. Thus the crab becomes indistinguishable from its environment. Are performances such as this directed by some intelligence which we cannot fathom? The need of this creature for protection against its predators existed, and in some way, the decorator crab was able to provide itself with the structures necessary to accomplish intricate and involved procedures for fashioning this disguise.

Our mythical travelers are overwhelmed with wonder. One can imagine that they are muttering, "I can scarcely believe what I have experienced, but it is real. I have wandered through these many encounters and have witnessed the most marvelous sequences of adaptive structure and form, and the seemingly intelligent use of such adaptive structures to perform the most complicated of actions, and the built-in genetic means of passing them on to their progeny." As our travelers proceed forward in cosmic time, they must realize that what they have seen, so far, is only a small drop in the total pool of evolutionary history.

Drawing of a decorator (spider) crab, which has placed samples of the environment on its back as a means of camouflage.

Part 3: LIFE PROLIFERATES ON PLANET EARTH

Chapter 18: FROM SPIDERS TO DINOSAURS

Time is quantitative in that it can be measured in terms of space and motion. Time is an essence. It appears, is with us for an infinitesimal moment, and departs, never to return. And vast quantities of time, in association with the other components of the universe and natural forces, have always been a part of the creation of Earth and its evolutionary processes.

Life, according to fossil findings, did not appear on Earth until the planet had reached the age of 1.2 billion years. This vast period of time was necessary just to prepare the planet for the advent of life. It required approximately another 2.6 billion years before these first living forms became capable of producing a sufficient number of hereditary diversifications to result in multicellular soft-bodied marine forms. Fossil remains of these creatures have been dated to 800 million years ago. At some time during the next 200 million years there appeared the spiders, insects, and fishes.

Spiders share a common ancestor with the crustaceans, particularly with the crabs. There is, indeed, a group of crabs known as the spider crabs which, though they are genetically unrelated to the spiders, are sufficiently similar to inspire the name. In both spider and spider crab, the thorax and the head are joined. The part of the spider crab containing the digestive tract is not segmented, and in this it differs from other crustaceans. Both have the same number of paired jointed appendages. The spiders have departed from their not-too-remote ancestry in that they are equipped with lungs, making them capable of existing in a non-watery environment. A few of the 15,000 species have diversified from egg-laying to producing their young alive.

Almost all species of spiders are equipped with a special organ called the spinneret, which produces a continuous threadlike substance that hardens on contact with the air. This is the spider silk with which they produce netlike structures of great architectural complexity and beauty. When webs are properly anchored and placed within the pathways of flying insects, they trap insects for the ultimate consumption of the waiting spider. It is not uncommon among spiders for the female to devour her more diminutive male mate after he has

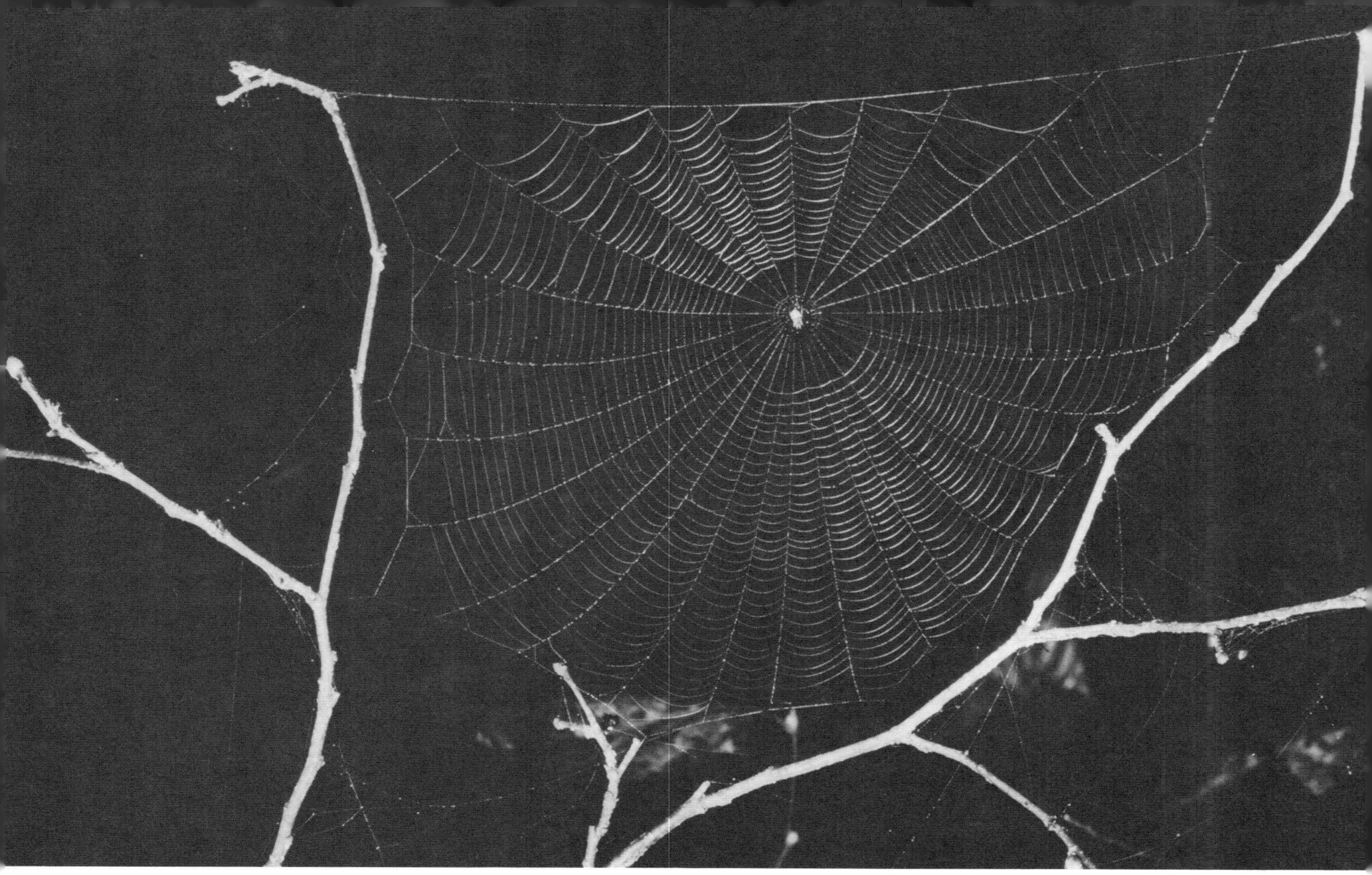

Having completed spinning its web, the spider awaits its unwary prey.

completed the act of fertilization. Thus the spiders, while retaining some of the structural and functional components of the crustaceans, have also diversified and adapted to a new environment, where lungs for breathing and webs for snaring food made it possible for them to cope and to survive.

A little farther along the steps of evolutionary time, there appeared a new family of living creatures called insects. Of these there have developed some 650,000 species. There are as many diversifications, adaptations, and functions within the creatures belonging to specific groupings in the family of insects as their environments have required. They are very important to the total ecology, and every major institution of learning and research has a well-developed department of entomology for the study of insects.

Insects have had both beneficial and destructive impacts upon Earth and upon its plant and animal inhabitants. Some insects, such as mosquitos, ticks, and lice, bear disease. They transfer infection from one animal host to another. Others devour or become parasitic upon certain plants. Insects can either partially or completely destroy the production potential of the agriculturist. Other insects, such as the termite ant, can totally destroy structures made of wood. And we have always had to contend with the irritating bites and stings of other insects.

But what about the benefits of the insect world? No one can deny the goodness of the honey bee, nor the many wonderful attributes of other pollinating insects. Without them, there would be little continuing plant fertility, and there would be little change in the genetic pattern of crop and floral plant life. And what about the larval form of the moth, the silkworm, which spins a cocoon with a fine, continuous thread? Without the silkworm, humans would not be able to unwind this thread and weave it into silk, a cloth of great beauty and texture.

The little fruit fly, with four chromosomes and a short reproductive cycle, does not know that it has provided, for the geneticist, a tool of great value. And the repetition of the phylogenetic past in the reproductive cycle of most insects, from the egg to the larva to the pupa (and in the case of the butterfly and the moths, to the chrysalis) has provided much to the scientists in the way of evolutionary interest.

Insects, probably the most diverse of all biological forms, have an important place in the evolutionary diversification of the world. Despite our lethal technology against them, the biological shift has been small indeed, and insects continue to play a dominant role. Thankful we must be that the magnificent beauties of the beetle and the butterfly, and of the moth and the firefly are still with us.

But what about our mythical travelers? Have they become weary? Are they unimpressed by this saga, by the spiders and the insects and their many contributions to the sequences of evolutionary progression? Indeed not, for they have learned, just as Sir Francis Bacon had, 400 years ago, that knowledge is power. Our travelers' knowledge gives them the desire to continue on their journey of searching, of sifting, and of winnowing from all that they have observed. For, all about them was the truth, and it merely needed to be recognized. So far, this has been the role of our mythical travelers, observing, integrating, and interrelating each and every segment of their experiences in order to distill from this combination the underlying truth and meaning. By now, they are aware that this is no idle journey through cosmic time and space. They are privileged to witness a series of events in which relationships have patterned the progression. They have become aware that each new evolutionary structure has carried with it some of its ancestral past. But there is more to observe.

Once again, as they stand and peer into the water, as they did a couple of billion years ago when the blue-green algae had just come into existence, our travelers witness yet another biological phenomenon, the minute creatures known as cephalochordates. Near the head of the body they are equipped with a spot-like structure which represents a primitive brain structure. Extending backward from this is a cord which is the precursor of the spinal cord. The earlier forms (crustaceans and insects and even earthworms) had brains and nerve cords too, but on the ventral or belly side. Here is a spinal cord on the dorsal or back side, which eventually will be enclosed and protected by bone. This primitive brain is capable of originating and transmitting impulses through the cord, and of receiving such impulses in return, so that favorable and protective reactions can result.

Scientists know them well. There is a small, twisting, squirming, extending, and withdrawing creature known as amphioxus, and a host of related creatures known as lancelets. There are also the later creatures, lampreys and hagfish. These are true vertebrates but without firm spinal cord coverings. Their mouths have no jaws; they sustain themselves by attaching their sucking

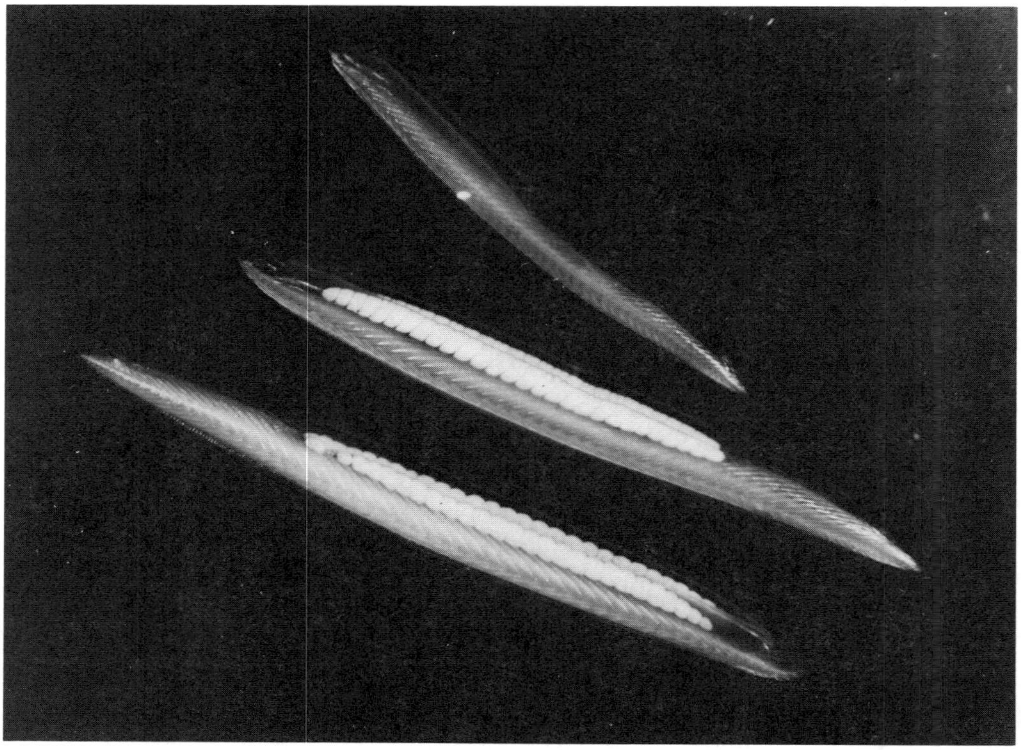

Photo of amphioxus. Note segments of muscles.

mouths to the surface of other fishes. The destructive effect of the lamprey on the lake trout population of the Great Lakes is well-known to both commercial and game fisherman.

Another step of progression in evolutionary history came with the advent of the sharks and of the rays. In these forms the spinal cord became covered with a semisoft, pliable cartilage giving protection to the underlying vital structure without limiting swimming motions. These creatures, especially the sharks, look like plausible precursors to the bony fishes, but they are not.

Everyone knows about the fishes, for they have become an important protein food source. Fishing provides pleasure and adventure for many people, from the youngster who sits on the bank and hopes to catch that big one, to the professor, the theologian, the engineer, and even the President, all of whom attempt to outguess the fish in trying to have them take the lure.

Fish have provided many opportunities for the scientist to study evolutionary trends among marine forms. To the evolutionist, fish represent an important episode in our evolutionary past, for they influenced many adaptive structural diversifications. Fish were the first of the vertebrates to have a bony covering for the spinal cord, the first of the begetters to provide for its begotten a structure which would ultimately make it possible for creatures to assume an upright posture. Fish were, also, the first animals to develop, on the undersurface of the body, two pairs of fins used for balance. These, ultimately, would become paired limbs. The fish developed scales, similar in function and structure to feathers in birds and hair in mammals. Fish acquired a unique structure

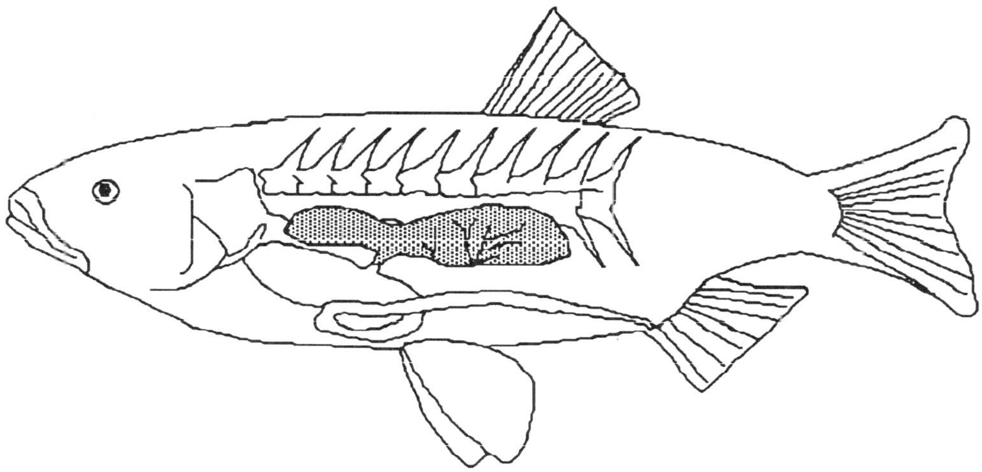

Diagram showing air or gas bladder of a fish. Note that the position of the air bladder is similar to that of lungs in air-breathing animals.

called the air bladder, or gas bladder, which allowed them to utilize atmospheric oxygen, and which could also be used (because the amount of air in the bladder could be regulated) to enable fish to suspend themselves in the water, at any depth. This adaptive structure, the air bladder, is important in another sense in that it provided the means for adaptive change. It developed into the lung, a structure which enabled subsequent creatures to breathe air. We are greatly indebted to these inhabitants of the oceans, lakes, and streams for their many contributions to our own remote, evolutionary past. And when we remove trout, bass, or perch from our hooks we should fondle them with care and be ever thankful.

But let us get back to the air bladder, for it provides a good example of evolutionary change, the reuse of structures for functions once discarded. Early in its evolution,the air bladder provided for utilizing atmospheric oxygen, allowing fish so equipped to live in stagnant water. This function gradually disappeared, and the air bladder was used only for depth control (buoyancy) in locomotion. Eventually, with the garfish, the original use was readopted. These fish come to the surface and expel old air from the bladder and fill it with new air, enabling them to remain underwater for long periods of time.

Another adaptive change in the air bladder came about with the advent of a species of fish known as the lungfish, in which the air bladder became paired, resembling primitive lungs. Along with this change, the fins become jointed, which enabled the lungfish to crawl on muddy bottoms and to dig a burrow to a depth of 18 inches, where during periods of drought, it could survive in a moist environment for periods of six months or longer. The primitive lungs made it possible to have a sufficient supply of oxygen in this somewhat dormant stage. When the rains came and the waters rose, the lungfish could again assume their stream and lake dwelling habits.

At this place in the chronology of evolutionary history, there developed a creature with jointed appendages for locomotion, with lungs for breathing air, with a bony skeleton for support, and with scales to protect the soft tissues. Such creatures might have poked their heads from the water to determine what land and air oriented existence could be like, and even to crawl a little farther and a little farther from their stream or lake habitats. Eventually, they discovered that they could have the best of both underwater and of land and air oriented existences. The way was paved for the advent of the amphibian.

Let us think a little about these amazing creatures, the salamanders, toads, and frogs. Although they live out of water, they must lay their eggs in watery surroundings for, after hatching, the young must live in the water for a period of time and go through the various stages of their phylogenetic ancestry. They are swimming creatures up to a point of development. As a remnant of their acquatic pasts, their skins must be kept moist, so they reside near or about water. The gills and some of the other marine structures are lost in this process of metamorphosis. Salamanders, unlike toads and frogs, do not lose their tails; but, like them, they develop four legs. All members of the amphibian family have the ability to regrow tail or leg structures if they have been amputated by predators or lost in some other trauma. Most fishes and amphibians lay tremendous numbers of eggs for, if they didn't, the loss due to predation would have a serious impact upon their survival as species.

Amphibians and fishes have been successful. Ever since the amphibians, the bony spine has been part of most future creatures in the evolutionary line. It is likewise with the lungs, with the finger-like toes of the frogs, the toads, and the salamanders, and with a life oriented to land and adapted to land and air.

The first group of distinctly land and air oriented creatures to come upon the scene was the reptiles. Their impact on evolutionary history was indeed tremendous. Reptiles include alligators and crocodiles, turtles, serpents, and dinosaurs. Representatives of all members of the reptilian branch are still with us except the dinosaurs, which became extinct some 65 million years ago. We know them only by their fossil remains.

Although to the casual observer the various species of the reptilian branch may seem to be diverse, they are in all structurally closely related. They are all cold-blooded, which means that they regulate their body temperatures by exposure to the sun or by withdrawal to the shade. Some of them, like crocodiles, immerse themselves partially in even-temperature stream waters. All members of this group have scales, although in some instances they are somewhat modified. They have fully developed lungs. Their extremities have the same bony structures as are found in mammals, even though these bones may be covered by flesh and skin to form flippers, as in the case of some turtles. In addition, all egg-laying reptiles lay their eggs on land, for their eggs have a semihard covering which protects the inner portion from drying out.

There are a few species of reptiles which have become viviparous and give birth to live young. Among the fishes (such as the grunion) there are instances of a cyclic reproductive urge, in response to moon, tides, and seasons. Among the amphibians and the reptiles these cycles are the rule.

Let us examine the reproductive strategy of a unique type of turtle whose homeland is along the coastal areas of Brazil. This turtle, after responding to cyclic influences (perhaps the length of the day, an indication of the seasons), becomes fertilized by her mate. She swims to a tiny island off the coast of southern Africa, 2,000 miles away. Upon arriving, she emerges from the water and flippers her way to a place on the sandy beach which the high tide will not reach. There, after removing sand to a depth of about 18 inches, she lays 200 or more eggs. After doing so, she replaces the sand and by mean of her flippers removes all external traces that she has ever been there. When the young hatch and emerge from their sandy nest, they will make their way to the sea. Those which survive predators swim to the coast of Brazil, from which their mother came. Some scientists are of the opinion that the genetic imprint upon the reproductive urges of these turtles dates from before the time, over 200 million years ago, when South America drifted away from Africa. May not this type of travel for reproductive purposes be related to that which is undertaken by salmon, eels, and shad, when they return to the rivers of their breeding? It is indeed interesting to speculate on these phenomena.

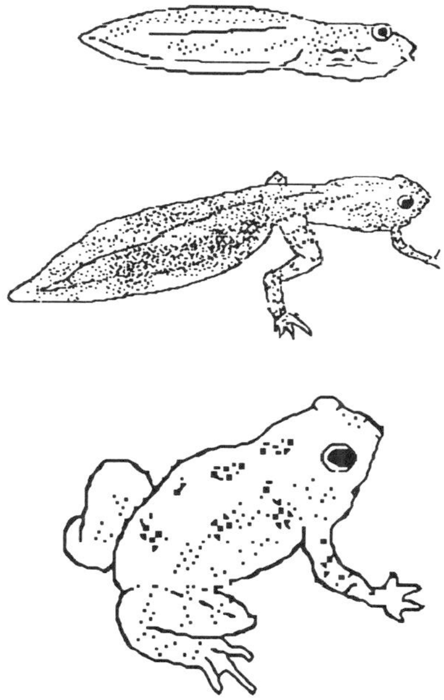

Metamorphoses of a frog showing three phases of its gradual development from fertilized egg to adult.

There are certain types of lizards (the whiptailed lizard of New Mexico is one) which can lay eggs which hatch without having been fertilized. This is known as parthenogenesis, or virgin birth. It produces only female offspring. (The term parthenogenesis is formed from the words parthos – virgin, and genesis – birth or origin. In the 5th century B.C., the Greeks built a beautiful Doric temple on the site of the Acropolis in Athens. They dedicated it to the virgin goddess Athena, and named it the Parthenon.) The cells of the offspring have the same number of paired chromosomes, identical in every way to those of the mother. The offspring are, therefore, clones of their mother, and the only variations come by the somewhat remote and rare process of mutation.

In the January 1978 issue of *Natural History* there is an article titled "Virgin Birth," written by Walter J. Cole. He describes in detail the processes involved. In experiments conducted in the laboratory of the American Museum of Natural History, four generations of whiptailed lizards have been reared, none of whom has ever had any contact with a male mate. They continue to produce eggs which hatch only into female offspring, all of which have the same genetic composition as their mothers.

Virgin birth is an amazing event indeed, but it is not new to evolutionary history. Such reproduction occurs among certain protozoans and among algae and other multicellular animals such as water fleas and aphids. The whiptailed lizard is not alone in parthenogenesis even among the other vertebrates, for there are some fishes capable of this mode of reproduction. Both sexual and asexual reproduction serve the species.

There are many species of lizards, such as chameleons, gekkos, and skinks. These are harmless, beautiful little creatures who sun themselves, and whose males puff out their highly colored throats and at the same time issue faint but audible mating calls. The females rush to the source of such a call, involve themselves in a momentary mating embrace, and thus the reproductive cycle goes on.

So far, our travelers have experienced encounters only with creatures of limited size. Now, they prepare themselves for the enormity of the dinosaur.

What thoughts must be going through the travelers' minds? How can they grasp the magnificence, the perfection, the beauty of the orderly sequences, the awesomeness of all that they have experienced so far? How can they reconcile all of this with mere mechanism, mere happenstance? On the other hand, if all of these encounters were preordained and predesigned prior to execution, what is the nature of the power and of the force behind it all?

Part 3: LIFE PROLIFERATES ON PLANET EARTH

Chapter 19: FROM DINOSAURS TO MAMMALS

In the latter part of the 19th century, Roy Chapman Andrews, first president of the American Museum of Natural History, gathered together a few of his adventurous friends and planned and executed a trip into the Gobi Desert. The purpose of this trip was to search for fossils or other biological remnants that might be added to our knowledge of our ancestral past. Among other things, three large eggs were found, one of which contained the fossil remains of an intact dinosaur embryo. These eggs are, at present, on display at the American Museum of Natural History. Prior to that time, dinosaurs were known only by their fossilized bones and by footprints.

The last of these creatures became extinct about 65 million years ago. The earliest of them came into existence about 200 million years ago, according to well-documented, radiometrically dated bones. The earliest of the dinosaurs to appear upon the scene seem to have been the largest, and as their dominance waned they seem to have become smaller before becoming extinct. The largest dinosaurs weighed about 200 tons and were about 75 feet in length, while the smallest were no larger than an ordinary pigeon. Until somewhat recently, they were classified as poikilothermic, or cold-blooded creatures who regulate their body temperatures from their environment, chiefly by the degree of exposure to sunlight or shade. However, more recently there has arisen a lively controversy about this point. In an article titled "Second Thought about Dinosaurs," published in the May 15, 1978 issue of *Newsweek*, Robert Baker, paleontologist at Johns Hopkins University, asserted that dinosaurs were not the slow, lumbering, basking creatures that we have been taught to believe, but they possessed sophisticated internal systems capable of regulating and keeping their temperatures constant.

Footprints such as those found in Dinosaur Park in Connecticut and near Billings, Montana indicate that at least some of the dinosaur species stood upright. It has been argued that the dentition of the dinosaurs made these creatures capable of consuming vast quantities of food. Fossil findings in Alberta, Canada, where the bones of prey and dinosaur captor were found together, indicate that at least this dinosaur had a prodigious appetite.

John Ostrom, in an article titled "Dinosaurs," published in the August 1978 issue of the *National Geographic*, has given us much to ponder concerning this problem. His fossil findings near Billings, Montana, enabled him to reconstruct the skeletal structure of a dinosaur much smaller in size than those which cause us to be humbled when we stand in their presence at museums.

Ostrom's dinosaur was also long-necked and long-tailed. It had two pairs of extremities. The front extremities had certain wrist bone structures which would have enabled these arm-like appendages to hold and to transport food to the mouth. The rear extremities were longer, heavier, and in many ways similar to those of mammals. They were three-toed and equipped with grasping claws. The dentition of these creatures included canines, indicating that they were carnivorous. And Dr. Ostrom concluded that the legs might have been capable of inflicting forceful blows upon a prey. Once the prey was down, the legs might have been capable of tearing flesh by means of grasping

Excavation of river sedimentation in the 140 million year old Morrison Formation at Dinosaur National Monument reveals fossil bones of fourteen species of dinosaurs.

claws. Footprints of comparable size in Dinosaur Park in Connecticut were grouped together in such a manner as to indicate that perhaps such dinosaurs might have hunted in packs.

Could it be that the first dinosaurs, those enormous herbivorous creatures, were indeed cold-blooded, and that they became warm-blooded as adaptive structures developed which enabled them to hunt, strike down prey, rip it apart, and consume it? There are many pro and cons to this controversy. And the difficulties of taking a positive stand one way or the other are indeed great – primarily because the fossil remains of these creatures which roamed Earth during the age of the dinosaurs are scant. It is likewise not known why the dinosaurs became extinct. Some people have linked the extinction to the climatic changes which might have occurred as a result of the continental drift, at the time when the north and south American continents joined. Others have contended that there was an exploding star or asteroid which wandered too close to the Earth's surface, causing tremendous, destructive temperature changes. This latter seems untenable, because all other forms of life would have also been destroyed at that time, and we would not have our living descendents of these reptiles.

It is known that the bones of the dinosaurs were hollow, giving a degree of lightness such as that which the bones of birds possess, and that there is great similarity between the structural long bones of this dinosaur and the legs of the ostrich (which, as is well known, can inflict blows with tremendous force). It is also known that dinosaurs, at least those which appeared within the last few million years of the age of the dinosaurs, could plane through the air for short distances. But, as in all transitional evolutionary changes, there always has been and always will be some degree of overlap.

For example, in 1977 James Jensen of Brigham Young University

Lower Jurassic period fossil marine crocodile with gizzard stones. Found in southern Germany.

96 found, in 140 million-year-old sediment, a thigh bone which appeared to be that of a bird. But this is how theories of evolutionary progression can be disturbed. One thigh bone. Could this indicate that there was a bird dinosaur long before it is ordinarily thought that birds came into existence? Further evidence will be needed to prove or disprove this point.

Let us turn to a pigeon-sized, long-tailed, long-necked, cylindrical-bodied, feathered, three-toed creature known as archaeopteryx. This creature had the structure of a dinosaur, but its scales had become feathered and it had cone-shaped teeth. No fossil remains of any bird since archaeopteryx have had teeth. It must have been warm-blooded, for a body covered by feathers could not readily absorb heat from its environment. The skeletal structures of the breast had assumed the shape of the well-known wishbone, in order that wings could be attached, and the bones had become hollow for lightness. All of these structural adaptations made it possible for this amazing creature to propel itself,

Archaeopteryx, the oldest known fossil bird, dates from the middle Jurassic period, about 150 million years ago.

not by walking, not by swimming, and not by to and fro motion of its tail, but by means of flying through the air with feathered wings.

In 1861 in Bavaria, Germany, the first fossil form of this three-toed, lizard-tailed, feathered, winged creature was found. These fossils of an incomplete skeleton are now in the British Museum in London. A few years later, an almost complete fossil skeleton of an archaeopteryx was found in a nearby quarry. These fossil remains can be seen in the Museum of West Berlin.

Archaeopteryx fossils had cone-shaped, almost canine teeth. Note that all birds which had dentition, originally, are now extinct. The archaeopteryx was the beginning of a new living form. There have now developed over 14,000 species of birds composed of adaptive structures to meet the special needs of their surroundings.

There are waders such as flamingos, great blue herons, ibis, and cranes; and there are swimmers such as ducks, geese, swans, and roseate spoonbills. There is the graceful flight and dive of the pelican, and there is the nectar-gathering of the hummingbird, which can fly in reverse and hover. There are songsters, vast numbers of them, each with its beautiful, musical song to identify it, and there are multicolored parrots, and cockatoos. There are predators which soar until they spot their prey and then dive bomb it. One could go on and on. What would the earthly domain be without the birds to delight the disciples of Audubon?

By this time, our mythical travelers are aware that each new biological form has been able to take something from its ancestral past, and to alter and change it in such a way that some of these adaptations made it possible for them to survive in their changing environments. But also, the adaptation was able to become part of the genetic patterns of the future.

Thus it was that the fishes provided the spinal structure which would ultimately become the support for assuming the erect position. The swim bladder of the fishes became the lungs necessary for land adaptation and air breathing. The scales of lizards became the feathers for flight and were ultimately modified to form the hair of the mammals. This has been the pattern from the beginning – some of the old becoming a part of the new. And this is what our travelers learned and will continue to learn until the end of their journey. It seems to them, as the element of time has been consumed, the progression of biological evolution toward more and more complexity has increased. The creatures have met the demands of the marine environment, then of the land, and most recently of the air.

And all of them carried with them something in the way of structure, form, and function which was possessed by their ancestors – modified, however, in order that they could become more diverse. This is the amazing thing about biological evolution. It has been a positive increment of biological progression from the simple to the more complex, and our travelers cannot abandon the thought that something which they cannot quite comprehend has been working in a most mysterious manner.

Part 3: LIFE PROLIFERATES ON PLANET EARTH

Chapter 20: THE AMAZING MAMMALS

There is an interesting, oft-told anecdote about Michelangelo which goes something like this. The block of marble stood before him; it had been there for some time. Michelangelo would occasionally come, sit before it, examine it from all sides, and go away. Then the time came for him to begin his work. He assembled his tools and began to chip away. Each day as he chipped he noticed that a young man stood silently nearby. His visitor would appear every morning and remain silently until Michelangelo had finished his daily task. It was only after Michelangelo had finished the beautiful sculpture of David that the daily visitor approached and said to him, "How did you know he was in there?" Only Michelangelo had known that David was imprisoned in this rough, stony mass, and only Michelangelo, utilizing to the fullest his superb art and skill, had been able to liberate him, so that his beauty and magnificence could be shared by all.

So it has been with our scientists who have directed their efforts at probing into the secrets and the mysteries of the universe in an attempt to try to find the answers to our origin and being. They know full well that the answers are there, but like Michelangelo's David, the answers are hidden from view. Also like Michelangelo, scientists have faith that by developing and utilizing specific, advanced technology the darkness that obscures their quest will be removed and the various facets of truth will eventually emerge.

Unlike our 20th century scientists, our travelers have been able only to observe and to record. They have been unable to advance further along the pathways of evolutionary times than the chronology of these pages of evolutionary history has permitted them to go. They have witnessed the first 4,535,000,000 years of Earth's evolutionary history and stand on the brink of entering the last 65 million years of Earth's existence. They are about to leave those marvelous creatures, the crocodiles, the lizards, the serpents, the now-extinct dinosaurs and the archaeopteryx, all of whom have contributed so much to evolutionary sequence and progress. A new array of creatures, the amazing mammals, is about to appear before them.

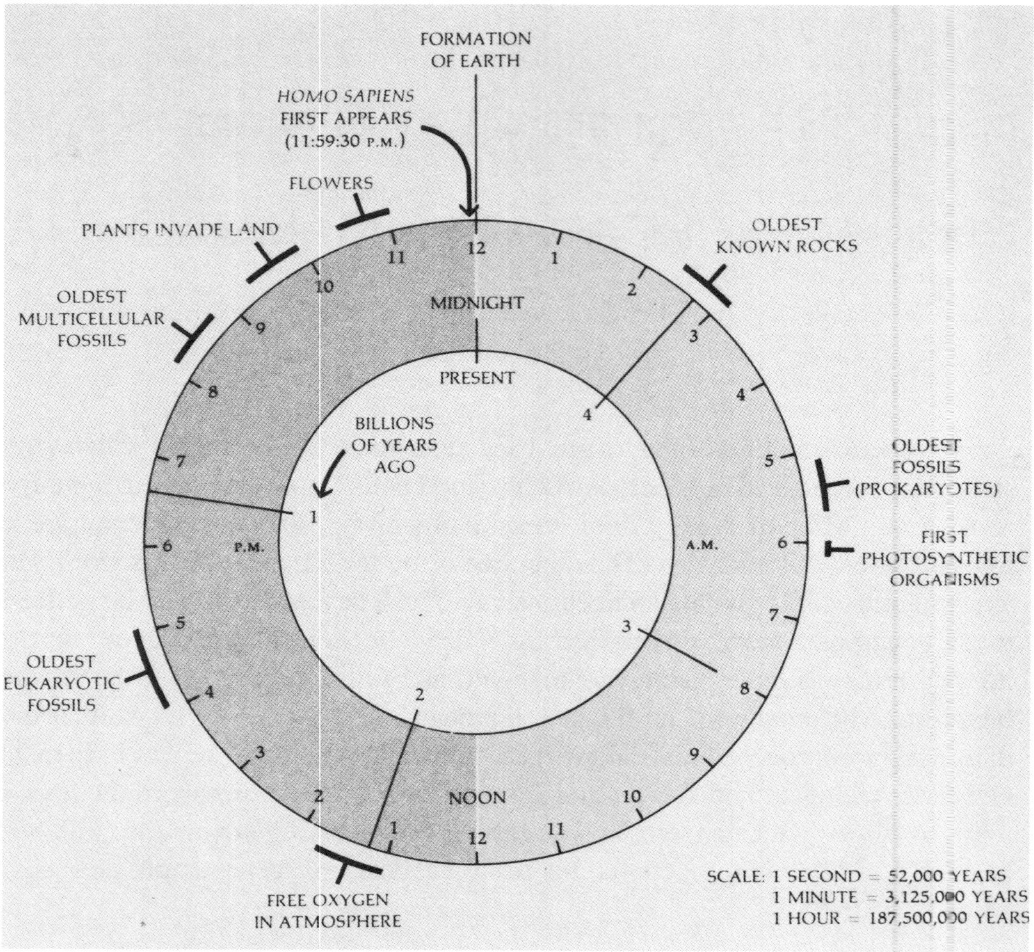

A schematic representation of time periods from the formation of Earth to the present as indicated on a twenty-four hour clockface. Note that humans appear in the last half-minute.

This period of biological ancestral history is important in that humans belong to this great branch of mammals. However, the immediate progenitors of Homo sapiens will not begin to make their appearance on Earth until about the last 3 million years. When compared to the total age of Earth, humankind has had its habitat thereon for only a brief moment of cosmic time. It has taken a long, long period of evolutionary change in order to permit the ultimate goal of creative, biological perfection to appear upon the scene. We shall try to learn in what way the early mammals contributed toward this steady advance along the pathway of evolutionary progression which has made possible the achievement of this ultimate goal.

Mammals are living creatures capable of nursing their young. For warmth and protection, their bodies are covered with hair, rather than scales or feathers. Like all other higher biological forms, their reproduction begins with the fertilized egg. However, with a few exceptions, the fertilized egg remains within the body of the female, where it grows and develops, until, upon reaching a certain degree of maturity, it is expelled to the external environment by the birth process. Most mammalian infants must, for various periods of time, be kept warm, be provided with nourishment, and be protected until they are ready for independent existence.

So far, evolutionary progression has taught us that one group of a species does not suddenly and immediately become capable of assuming advanced, functional performance roles. The members of the reptilian orders had to go through transitional stages prior to becoming advanced mammals. What then were these transitional stages and what kinds of creatures were involved in them?

There were a few such intermediate forms which are worthy of note. They possessed characteristics reminiscent of their ancestral pasts while at the same time possessing structures and functions characteristic of mammals. There is, for example, a wonderful little creature known as the duck-billed platypus, found only in southern Australia and on the island of Tasmania. To judge from appearances, this 14"-long, semiaquatic animal has the bill of a bird – a flattened sort of duck bill – the webbed feet of the lizards and the frogs, claws on its toes similar to those found on the late dinosaurs and on the feet of birds, and some of the internal structures of the reptilians. Like the mammals, it also has fur on its body, and well-developed, paired breast structures – but without teats – and a broad, flappy tail like a beaver's, which can fold over onto the abdomen.

Appearances, however, are not an accurate guide to evolutionary relationships. The bill, covered with skin, is not structurally like a bird's, and in no way evidences a genetic relationship to birds. The webbed feet are more like a beaver's – simple adaptations for swimming, and not evidence of descent from lizards. Likewise with the other structures. They are present because they are necessary for performing functions which are essential to the well-being of the platypus and to its offspring. For example, the claws can dig a burrow which the female platypus constructs, usually alongside a stream or lake. One opening is above the water level while the other leads into the water. While on land, the platypus can gather leaves, moss, and grasses to pad the underground nest. Upon entering the water from the other opening the platypus can use its duck bill to stir the bottom in order to liberate worms, snails, and insects for nourishment. The thick brownish-gray hair keeps this creature warm.

The platypus lacks one of the characteristics of the advanced mammals, however, in that it has not yet developed the ability to nurture its young within the body of the female. It is an egg layer. It produces one to three eggs on the padded nest of its burrow. The hatchlings do not have bills, but are equipped with a sucking mouth somewhat similar to that of the lamprey (though, again, no evolutionary, genetic connection can be deduced from the apparent similarities of form or function). The young grasp the soft tissue overlying the breast and suckle the milk – not through teats, for the platypus has not acquired this characteristic – but through pores. The flappy tail of the platypus folds over the abdomen and holds the young hatchlings in place. The young of the platypus have teeth similar to those which were possessed by the late dinosaurs and the earliest birds. What a wonderful creature is this platypus! It tells a story all its own.

The spiny anteater is another most interesting egg-laying mammal. Like the platypus, the anteater is native to southern Australia and the island of Tasmania, but it is also found on the islands of New Guinea. Like the platypus, it has a beaked mouth and claws upon its toes with which it can dig shallow burrows. It can tear apart the structure of an ant hill or mound and reveal the concealed insects to be feasted upon. The flappy tail the platypus uses to hold its young to the milk pores was not good enough for this creature. It did not need the tail for guidance in swimming, as did the platypus. It developed a pouch in which to place its hatchlings so they could be kept warm and nurse while they matured.

This pouch, a signal contribution to evolutionary progression, was developed fully in another interesting group of mammals, the marsupials. This group includes kangaroos, wallabies, wombats, koala bears, and a few wolflike creatures, all native to Australia. The only North American representative of the marsupials is the opossum, ranging in habitat from Virginia to Florida and westward to the Great Lakes area. All of these mammals are equipped with teats. The eggs are fertilized within the body of the female and develop in the mother's womb to a fetal stage. Then they are born and make their way into the pouch immediately, where they suckle and remain for a period of time until they mature.

In the case of the kangaroo, the first nursing occurs on the lower pair of teats, for these breasts produce a certain nutrient formula which is adapted to this particular age of development. Only as they grow larger and older do young kangaroos ascend to the upper teats. While nursing this young one, the mother conceives another embryo. If anything happens to the one to which she has already given birth, the replacement is on the way.

The kangaroo can partially support itself on its long, strong tail, which gives it a semierect posture. The forelimbs have been modified to perform arm-like functions. One only has to witness the comical boxing antics of these creatures to see how arm-like these limbs are! Both tail and the arm-like forelimbs are reminiscent of similar structures and functions in the dinosaurs, but once again the similarity is apparent, not genetic.

The American marsupial, the opossum, developed structures which let it assume behavioral patterns that were new for mammals – living in trees. Enlarged, curved claws let this creature climb trees; the hairless, prehensile tail, curved over branches, allowed it to hang securely upside down. This combination of tail and limbs enabled it, first among the mammals, to live an arboreal existence. Hair, flappy tails, pouches, breasts, reproductive structures within the body of the female, forelimbs capable of handling and fondling the young, tails, and climbing claws – all brought tremendous advances along the pathways of evolutionary progress.

Mammals would not find it necessary, as their reptilian ancestors had, to deposit eggs upon the land, where both predators and natural catastrophes take their toll. Marsupials would be able to have complete control over their offspring from the moment of fertilization through intrauterine development to the fetal stage, and from then on within their own pouches.

From now on, all new female mammalian forms would be capable of developing fertilized eggs within their own bodies. One great difficulty remained in that the continuation of life was still too hazardous for these immature birthlings. It was indeed better than the external egg, but still not good enough. The evolutionary process was not to be stopped here, so another remarkable advance came into being – the development of the placenta.

In placental mammals, soon after the fertilized egg begins to divide, some cells are set aside and develop into the placenta. This becomes the structure by means of which the developing egg can be attached to the wall of the uterus and receive nourishment. These special cells grow in number. Some of them grow into a great network of arteries, veins, and capillaries which provide a means of absorbing nutrient material from the maternal bloodstream and a means of discarding waste from the developing embryo. Eventually the umbilical cord, a structure containing arteries and veins, makes the connection between the blood circulation of the placenta and the embryo. Now the mammalian mother has within her body structures which enable her to bring her young to a greater degree of maturity before giving birth. Since the placenta, umbilical cord, and membranes which surround the developing fertilized egg are all a part of the product of conception, they too are expelled following the birth of the young. The placenta is thus commonly termed "afterbirth."

Each new fertilized egg must develop a new placenta, a new umbilical cord, and new membranes to surround the developing embryo. Even though such offspring are far more mature at the time of birth than those of the marsupials, before they can be on their own, they still require a period of care for warmth, nourishment, and protection. But what is more, this period of time provides an interlude of great importance, for by nursing, touching and fondling, great bonds of attachment and love can be established between a mother and her offspring.

All mammalian mothers and offspring exhibit these bonding attachments. Watch your mother dog or cat tenderly care for her newborns, and observe the length to which she will go in order to protect them from danger. Offspring which survive and grow to maturity will re-enact the role of their parents by reproducing other generations of their kind. They also will love and care for them in the same fashion.

There are many orders of mammals. They vary in size from the common mouse to the great whales. The same Dr. Roy Chapman Andrews who mounted three expeditions to Mongolia in the 1920's and discovered the three dinosaur eggs also discovered, in Mongolia, a giant mammoth weighing thousands of pounds. This mammoth had been mired in the muddy bottom of a lake from which it had been unable to extract itself. Its pelvis contained an intact embryo.

There are mammals that burrow to form a habitat for warmth, protection, and winter hibernation. There are those that can survive the severity of the arctic and subarctic climates and those that can cope with the torrid climates of the tropics. And there are those amazing mammals of the sea, the dolphins, porpoises, and the manatees or sea cows, and whales.

These creatures have all of the reproductive structures of an advanced mammal, and they develop a placenta and an umbilical cord. They give birth to their young in a fairly mature state and they nurse them. Since these creatures depend upon atmospheric oxygen for their survival, and since the newborn must initiate the breathing process from the atmosphere, one of the first acts the mother dolphin, manatee, or whale must perform is to immediately sever the umbilical cord and nose her offspring to the surface. Otherwise, it will perish from oxygen deprivation.

One interesting mammal of evolutionary significance is the armadillo. Its body is protected by a segmented, shell-like structure. This is the first of the known mammalian creatures to develop more than one offspring from a single fertilized egg. It is the rule for the armadillo to develop and deliver identical quadruplets from one egg. Other mammalian creatures do this occasionally, but usually only to the extent of identical twins. It is unusual for triplets or quadruplets to develop from one egg.

Most mammalian creatures have found it possible to cope with their surroundings. They have had imprinted upon their inheritance those characteristics necessary for their survival. Thus it is that, in this our 20th century, we have living representatives of much of this period of our ancestral past. It is also true that they are the most recently evolved, and the processes of evolutionary change, including extinction, have not had as much time to work on the mammals as on other, older animal groups.

Our mythical travelers are unaware that these evolutionary successions have been bringing them nearer to that important group of mammals, the primates. Representatives of this group include monkeys, baboons, gorillas, orangutans, chimpanzees, and humans. Since the primates have all reached the most advanced evolutionary development our mythical travelers have encountered, we shall study them in greater detail and try to understand how these creatures, too, have related to humankind's ancestral past.

Part 3: LIFE PROLIFERATES ON PLANET EARTH

Chapter 21: THE ROLE OF THE PRIMATES

As our mythical travelers once again move forward, the horizon is undergoing continuous changes. They leave behind what has disappeared from view and become acutely aware of the new which is constantly appearing. They find themselves entering the world of the primates.What, indeed, is a primate? The term is taken from the Latin word "primatus," meaning the first or highest above all others. Within the context of these writings, the word primate refers to those groups of mammals which have reached the highest phase of evolutionary development up until now. Humans occupy the top rung on the ladder of biological creation – they are primates above all other primates.

How did it come about that humans acquired the superb structure, form, and function which make people so much more capable of diversity than any other member of the animal kingdom? Were these acquisitions the result of happenstance? Indeed, they were not. For the biological form to become human required the same evolutionary pattern throughout the ages of the primates as was present in all evolution prior to their appearance.

On constantly changing planet Earth, creatures which cannot cope with changes in the environment do not survive. It was necessary that variations in structure, form, and function develop in order to meet new environmental challenges, and to become a part of the hereditary patterns of evolutionary progression.

We recall that approximately 70 million years passed from the appearances of the first prosimians until the creation of humans. We must be aware that this was not a direct line of descent. It was more like that of a tree having a common rootstock with many branchings, with one such branch leading to the ultimate goal.Go back, then, about 70 million years as we attempt to trace some of the important evolutionary pathways along which our travelers must wander.

Somewhere near to the time when the last of the dinosaurs were becoming extinct, the marsupial opossum had just taken to the trees. Its short, five-digited extremities equipped with claws enabled it to do some vertical climbing and to clumsily find its way along the larger branches of the trees with its long,

Humans are among the least specialized of the many and diverse mammals.

whip-like tail providing an additional means of transport and support. This creature no longer needed to scurry about to guard and protect land-laid eggs. It carried structures necessary for the fertilized egg to develop to a certain embryonic stage within its body, at which point, by means of a pouch, the offspring could be nourished and protected until they were ready for independent survival.

However, there were certain drawbacks to tree dwelling. The opossum must hold close to the trunk or to the limbs of the tree because its extremities are not long enough to permit otherwise. It can not raise itself far enough to detect predators readily. It can not easily reach choice fruits or buds at the ends of branches. Finally, it does not possess the agility needed to swing speedily or freely from one branch to another.

It remained for another, contemporary line of development, that of the prosimians, to begin the series of structural changes which made them fully adapted to tree living. The bodies of the prosimians became somewhat more slender, their tails grew a little longer and developed a highly sensitive tip, the heads became round and the face became flattened, the mouth extended somewhat forward like that of a dog, but the eyes began to move from the sides of the head toward the front. The extremities became longer, the five digits on each extremity lost their webbing and also became longer, and since claws now were unnecessary for the act of burrowing, these useless structures began to disappear and were gradually replaced by nails.

These creatures did not have the marsupial pouch, but were equipped with internal genital structures allowing them to carry the young to a higher degree of maturity before giving birth. There were breasts fully equipped with teats for the young to suckle.

Now, it would seem that these creatures had reached the ultimate in structural modification in order to meet all that tree-living required. They could see more directly forward; they could increase their range of vision by raising themselves up from their trees of branches, supported on their longer extremities; they could more easily grasp the points of support, and by the utilization of their tails and one of their upper extremities they could swing from branch to branch. And, they could do all of these things with greater speed and agility.

At this stage of evolution, the great toe on each extremity was directly in line with all of the other toes, so that grasping had to be accomplished by simple bending of all five fingers or toes. Locomotion was quadrupedal. As evolutionary time progressed and newer simian types came into being, there were progressive improvements on many of these structural and functional variations, making such creatures even better adapted to living in trees.

Now, we must turn, briefly, to a geologic episode which had a profound influence upon primate evolution. It has been only within the last few decades that geologists have come to believe seriously that the great land masses of Earth were not always distributed as they are today, and that, about 200 million years ago, there was only one great land mass, which geologists call Pangaea.

At some unknown time, this land mass began to tear apart. The large components, at first, drifted away from each other, and then reunited into the present configuration – North and South America united to form the western continent, and the several components of Europe, Africa, India, and Asia united to form the eastern continents. These forceful contacts caused great mountain ranges to arise from the sea, most notably the Himalayas – where the Indian subcontinent met the mass of Asia – and the Alps. The eastern contour of South America and the western contour of Africa seem to match, like pieces in a jigsaw puzzle. This readily suggests that Africa and South America were once one. Magnetic soundings of the ocean floor presented the final proof that the continents have, indeed, been drifting apart, and still are.

What was the impact of this geological circumstance on primate evolution? When the break occurred and the land mass known as South America began to drift westward, it carried with it the prosimians of Africa at the time of its separation. This has provided one of the great examples of how changing environment has altered the course of evolutionary progression. For example, the early primates have always been indigenous to the tropics and sub-tropics. And while some areas in Central America and Northern South America have remained largely forested, these same climatic zones in Africa have changed in their flora, having in some areas lost their trees and become grasslands or savannah. It is impressive to note that in the Americas, the simian types have not progressed beyond the tree-dwellers. There are many variations among the simians of the Americas, including spider monkeys, capuchins, and the largest, known as the howler monkey. (One needs only to be present at dusk in the jungles of southern Mexico or northern Guatemala in order to appreciated why this designation has been given. When large numbers of these creatures decide to sing their nightly chant, it is terrifying indeed.)

In these tropical and subtropical areas of the Americas in which primates abound, there has been insufficient environmental change to either require or make desirable variations in structure or function among the primates. They have been able to cope well with the environment.

Not so in Africa. New factors entered into the picture. It is thought that when large areas of previously forested land became savannah or grasslands, the primates found it necessary, at least part of the time, to venture from their tree habitations and to wander into the grasslands in search of food. The smallest of the apes, the gibbon, had adapted well to trees which were not quite so densely foliated as the trees of the new world. It was the first of the swinging apes. This act, called brachiation, is performed with the most superb grace, accuracy, and agility, using one arm for suspension and the other three hands – for this is what they really are, hands, and not just feet – can be used for grasping tender shoots and other available food stuffs. These antics can also be employed to scurry quickly from unwelcome territory or from the danger of predators.

Orangutans, next in succession among the ape-like creatures, have bodies that are somewhat larger and bulkier. They also use the swinging technique, but are not quite as agile as the gibbon. Orangutans live mostly in the trees, but they occasionally venture into the savannah. However, for one of the largest of the anthropoid apes, the chimpanzee, wanderings into the grasslands are quite commonplace. The gorilla, in constrast, stays on the ground in dense forests. Among these creatures relative brain capacity has increased tremendously. They have inherited from their immediate past both stereoscopic and color vision, and an increase in the molar or grinding teeth to serve them for the ingestion of seeds, grains, fruits and leaves. The thumb has grown larger and can be opposed to the side of the index finger, which is necessary for the performance of more highly skilled hand procedures.

One other structural change occurred in the digits, or fingers and toes, of the forward extremities of the gorilla and of the chimpanzee. These digits became longer and jointed in a more sophisticated way so that they can be turned backwards and cover the undersurface to allow grasping. With these structural changes, the gorilla and chimpanzee can do what is commonly known as knuckle walking, and can grasp and carry with them grain or seed-containing straws as they knuckle-walk.

But what about the increased brain capacity of the skull? What purpose did this serve in the evolutionary sequences which led to the hominids and to humans? Certain well-directed researchers have significantly added to our knowledge and understanding of brain function as it relates to the primates.

Until recently, the late Dr. Harry Harlow was Professor and Director of the Primate Research Center at the University of Wisconsin-Madison. With his staff, Harlow planned and carried out research work directed at better understanding the interrelationships between the mother and her infant and the infant and its peers. Infant monkeys were taken from their mothers at birth and were isolated from any contact with other members of their species. They were given two artificial mothers, surrogates designed to resemble as closely as possible the actual mother. One of them was made of wire and was equipped with a bottle of milk containing a nipple. The other was made of soft terrycloth, but with no source of nourishment. When isolated individual monkey infants were placed within a cage with the two surrogate mothers they invariably chose to cling to and cuddle the soft terrycloth replica even though it gave no nourishment.

Such infants had to be artificially fed. When isolated infants grew to mature size and were placed in a cage with one of their peers, they exhibited no desire to have pleasurable social contacts. As a matter of fact, they stalked each other, and when contact was made they resorted to violent struggle. Even when left alone, mature monkeys which had been isolated at birth resorted to bizarre rocking movements and in some cases self-mutilation. There were no tendencies to sexual involvement by those mature monkeys who had been isolated at birth. Those who had been placed in contact with a moving surrogate mother were more curious and less violent than those who had been with a stationary one.

Isolated females, upon reaching maturity, were impregnated through artificial insemination. When their young were born such mothers did not seem to know how or want to fondle, nurse, or care for their young. In one generation, they seem to have lost this ability. Studies such as these which were conducted by Harry Harlow and Mary Freiburg leave little doubt as to the importance of maternal-infant contact and maternal-infant bonding. There are even some indications that the absence of such bonding may produce irreparable brain damage in the young.

112 Jane Goodall has devoted a great portion of her adult life to living in the African forest and studying the chimpanzee in its native habitat. She has revealed to us much about the life pattern of this most advanced of the anthropoid apes.

Another very interesting study was made in Borneo, Indonesia, by Birute Galdikas-Blindamour, who had spent the greater part of eight years among the orangutans. While these orange-colored apes for the most part spend their lives in the trees, they do venture to the flatlands for some of their activities. Galdikas learned that they are solitary animals and that the males and females remain separate except for the time of copulation. The social hierarchy is one of male dominance. In the absence of such a male, a female is dominant. The males fight violently over a choice mating partner. The females nurse and care for their young until they are about three years of age. They construct the nest high in the trees and there give birth to their young.

Throughout their eight year stay in Borneo, Galdikas and her husband had about them, in their quarters, five to seven young infant orangutans, and for about five years of this time their son Bin played with and grew up alongside of these maturing orangutan infants. Their play and other activities were for the most part similar, and Galdikas played with, fondled, and fed the young infant orangutans just as she did her son. There were many communications by gesture and touch. One bit of worry was that Bin seemed to have adopted the posture and arm swinging tendencies of the orangutans, but this quickly disappeared when the association no longer existed.

The first of the young orangutan infants which had been reared with Galdikas's family was returned to the forest upon reaching the age of six. Shortly following this, three young orangutan infants were found to have been killed. It was quickly learned that the six-year-old who had been reintroduced to the forest was the culprit. Galdikas and her husband concluded that these acts had been performed in fits of jealousy because the love and affection of Galdikas as surrogate mother had been dispensed to others. Acts of this kind are not a part of the orangutan's life pattern, and apparently were the result of intense emotional disturbance.

Tool use has been noted among orangutans. They will occasionally select a twig of proper length, break it from its branch, and use it for scratching the back. Branches are used as threatening weapons.

Without question, among the anthropoid apes, even as in many of the lower animal forms, there are means of communication which, in many instances, are obscure to humans. However, among the simians and the anthropoids, there are specific body extremity and facial gestures which, when coupled with a few sounds, indicate pleasure, fear, or danger.

Expression by any intricate means of vocalization, for the most part, does not exist. The famous Washoe and Sarah, two chimpanzees under the tutelage of Alan and Beatrice Gardner, demonstrated beyond any doubt that they can recognize the spoken word as applied to a given object and designate the object by the application of a given symbol. In addition, they can use symbols to construct simple word groupings to form sentences. Under the tutelage of the Gardners, Washoe and Sarah also learned how to use the American sign language – that which is used for the deaf – and they were able to learn over 150 utilization techniques.

Bernard Rench has reared a chimpanzee, Julia, and has taught her to engage in other acts of intelligence and of problem solving. He constructed six boxes; the cover of each was secured by a padlock. These boxes were placed in a row and the key to each preceding box was placed so that Julia could see it. The last of the six boxes contained the reward, a banana. Julia learned to unlock these successive boxes, and to claim her reward. In another act of problem solving, Julia was taught to push objects through complicated mazes in order to obtain food rewards. The chimpanzee performed these acts with degrees of accuracy comparable to those of human students; however, she proceeded more slowly.

From these and many other experiences with the primates, it seems beyond doubt that the intelligence, communication, and problem solving which is manifest in humans had its roots deep within our primate ancestry.

Part 4: HUMANS COME TO PLANET EARTH

Chapter 22: THE DARWINIAN HYPOTHESIS

In the year 1871 in the city of Chicago, a significant event occurred in the early morning hours. As she had been routinely accustomed to doing, Mrs. O'Leary was milking her cow. In the process, the animal moved and kicked over a lighted lantern. The flame within quickly ignited the straw which had been strewn over the floor of the cow shed. The rapidly spreading fire first destroyed the cow shed, then the flame, fanned and carried by the high wind, spread in an ever increasing circle, destroying building after building, until the tremendous conflagration ultimately reduced to ashes a large portion of the city of Chicago.

In this same year of 1871, another kind of fire was lighted. This was not intended to be a fire of destruction, but one of enlightenment. It was hoped it would grow in the minds of men to immense proportions. This fire was not to destroy a city but to wipe away the mountains of ignorance with which humans have been surrounded concerning their biological origins, their evolution, and their being. It was in this year that Charles Darwin published *The Descent of Man*.

The first sparks of biological evolution had preceded him. Jean Baptiste Lamarck much earlier expressed the view that evolution had proceeded along a more or less direct line of descent. Charles Darwin's grandfather, Erasmus Darwin, publicly expressed his views of a somewhat limited acceptance of the theory of evolutionary progression. Thomas Huxley, a contemporary of Charles Darwin, not only agreed with Darwin, but defended him on more than one occasion.

It was Darwin himself who observed, on his round-the-world ocean voyage, the many adaptations and diversifications of species. Many of these observations were published in *The Voyage of the Beagle*. Darwin decided that species had come into being by adaptation and diversification in response to environmental demands, and these ideas he published in *The Origin of Species By Means of Natural Selection*. Numerous other pamphlets and papers had described his many observations and their relationships to the evolutionary hypothesis. Then Darwin published his magnum opus *The Descent of Man*, in which he was the first to proclaim that humans were also part of this evolutionary progression.

In the early 1600s, not too long before Darwin's time, those who differed from prevailing beliefs and expressed their views accordingly were dealt with very harshly. They were subjected to the Inquisition, which was headed by the powerful Dominican Torquemada. The accused, when found guilty, had been subjected to the most diabolical persecution, torture, and in some cases death.

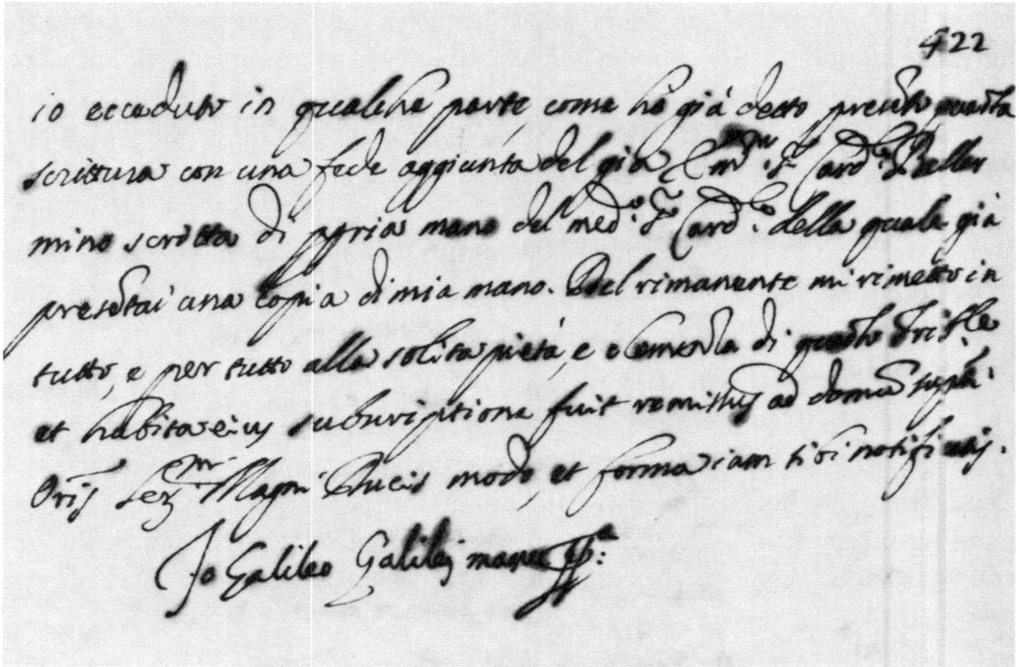

Galileo's recantal statement, written under the threat of torture, is kept in the Vatican Library.

Galileo was accused of heresy; he was condemned to the rack while listening to the agonizing cries of another individual who was having his sinews and his muscles torn one from another while his tormentors increased the stretching process. What had been Galileo's crime? He had simply stated certain ideas in which he honestly believed. The correctness of his observations have been substantiated by mathematical proof. He had simply stated that the Earth was not the center of the universe, that it was only a small planet subservient to the gravitational forces of our sun, about which it pursued a circular orbit requiring approximately 365 days. The theologians of the day would in no way tolerate the dissemination of knowledge such as this, so Galileo was forced to recant his beliefs under the threat of the torture of the rack and of excommunication.

Joseph Bronowsky, in his book *The Ascent of Man,* has reproduced in detail the trial of Galileo and the text of excommunication procedures. How horrible this all must have been. Had there continued to be such a climate against the rights of people to express freely those things in which they honestly believed, we would have continued to be surrounded by those mountains of ignorance which prevailed up to the time of Galileo. We would never be permitted to inquire into our origins. Bigots and their prejudices, like Torquemada and others, would not have permitted that. But, things had changed somewhat by the time Darwin arrived upon the scene. Courageous men like Bacon, Newton, and Spinoza had begun to deny the powerful inquisitors and their self-proclaimed right of silencing all those who expressed opposing views.

Spinoza, also, was brought before the Spanish Inquisition and was ex-

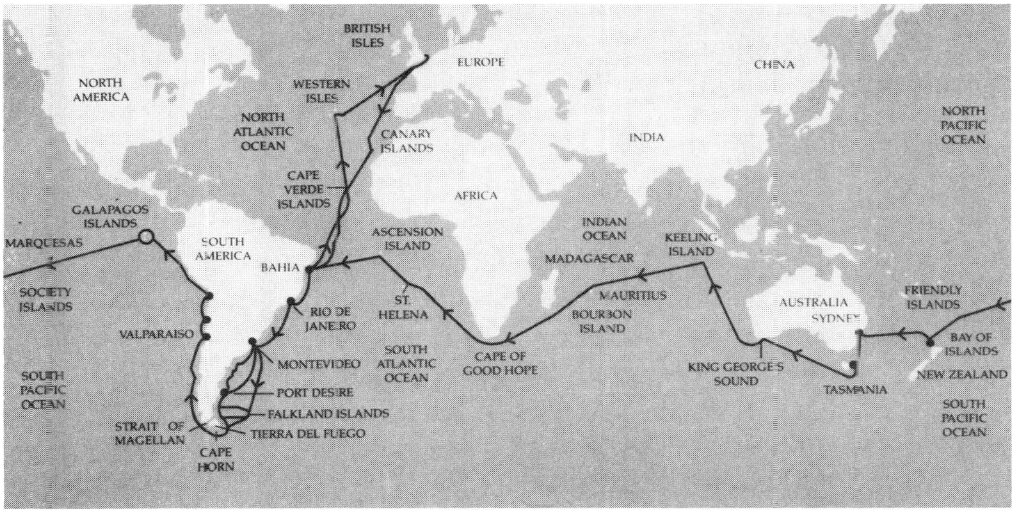

Chart of the voyage of Darwin's ship , the Beagle, on its trip around the world.

communicated and exiled to the Netherlands. He had already paved the way for freedom of thought and expression of ideas which were founded upon a firm belief. In 1651, Spinoza secretly published his thoughts about these concerns in his work titled *Tractatus Politico Theologicus*. The demand for his ideas, however, soon removed from this book the cloak of secrecy. That there was no heresy in spreading enlightenment and truth was an idea that was beginning to take hold.

Then Charles Darwin, courageous man that he was, published *The Descent of Man*, perhaps fully knowing the many controversies that it would ignite. After this publication, the fires of destruction were again lighted. Freedom of thought, of expression, and of enlightenment were once again being challenged. Those who would deny these privileges did not bring with them their weapons of physical torture, of excommunication, and eternal damnation to hell. They called forth other tortures which could be equally destructive – those of ignominy, ridicule, and the denial of the right to explore and to search for truth wherever it might be hidden.

Darwin merely stated that apes and man had been part of the same evolutionary process. A great cry went up. "Human beings descended from monkeys? How ridiculous! How absurd!" The wife of the Bishop of Worcester proclaimed, "If it be so that men had descended from monkeys, let it not be made generally known." Those who cried out so vociferously did so because of ignorance and prejudice, for they had not informed themselves of the true meaning of Charles Darwin's hypothesis.

Indeed, Darwin did not write that men were directly descended from monkeys, but that all living things had been a part of the continuum of evolutionary progression. This began with the creation of the first living substance and culminated, after vast periods of time, in the creation of humans. Along the way, many environmental changes occurred. Earth itself had been constantly changing and a series of new environmental patterns continuously placed new demands on living structure and form.

Creatures must, as a species, adapt and diversify in order to respond to these new conditions, or they would, as a species, perish. Evolutionary change happens on the level of whole populations through individual genetic change. Genetic change comes first, its usefulness for adaptation to environmental stresses comes second. Thus it is that, although we speak of an organism's changing or acquiring new structures, it is the species that changes, not the individual. The individual comes into being with genetic factors established. Sometimes this genetic endowment is of evolutionary significance. So, organisms which successfully met the requirements of new environmental circumstances through altered structure and function became the ones that were capable of survival. Alterations useful for survival continued to be reproduced, passed on with the other favorable, unfavorable, and neutral genetic characteristics to the individual members of the species.

In this way, the principle of the survival of the fittest became a part of evolutionary progression. In all of the time during which living matter has inhabited the Earth, up to and including our 20th century, it has been estimated by biologists that over 2 billion species of creatures have once lived or now live. Continuous environmental change is associated with all evolutionary progression, whether it be from sponges to spiders, or from apes to man, and has been with us since the beginning of biological forms.

Darwin also stated that this continuum consisted of a number of branchings. From these it was possible that one or more of a given species was able to survive and provide a progenitor stock for the various species types of the future. Thus it was that perhaps the mouse, lemur, or perhaps the tree sloth constituted the original mammal stock from which, after a long line of descent, there came first the prosimians, then the simians, and later on the anthropoid apes.

It was from such branchings that Darwin believed man descended, not directly, but by the ingenious mechanisms of adaptation, diversification, and modification of structure, form, and function to cope with environmental demand. Acceptance of the evolutionary hypothesis is still fought with high emotion to the extent that many fail to inform themselves of Charles Darwin's true meanings.

There are, unfortunately, those who are not only uninformed themselves, but, because of powers of oratory or of seeming religiosity, would attempt to keep all others equally uninformed. Let us think a little of what happens within the minds of casual observers standing before the primate cage in a city zoo, watching the monkeys perform their almost human activities. They leave, after a moment of observation, nothing more than amused by what has been seen. They do not realize that these creatures, also, experience problems, emotions, hunger and satiety, comfort and pain. People can fail to realize this even though they have watched a primate mother with her newborn offspring tenderly care for it, fondle and cuddle it, keep it warm, protect it from all who would endanger it, and feed and nourish it. Even though these traits are also common to humans, the observer departs with little or no sense of interrelationships. But primates are not just inferior humans. They are among the animals who constitute only one phase, only one of the sequential steps of evolutionary progression.

There must be further steps along the pathway of biological evolution; further adaptation, further diversification, and further branchings before humans can come upon the scene. We should be forever thankful that there were those such as Charles Darwin, Spinoza, Newton, and Bacon who, by their courage, abandoned the ideas of destruction and repression of thought which had chained the minds of men to one line of never-to-be-opposed dogma. Even today, there are those who would deny to humans the privileges of fully utilizing the minds with which they have been endowed.

Part 4: HUMANS COME TO PLANET EARTH

Chapter 23: THE APPROACHES TO MAN

Our premise is that humans have a common heritage with all other living things. No one can deny that everything that lives is composed of the basic unit of structure, the cell. It is also noteworthy that the cell can function singly, as manifest in the simplest of bacteria, or in closely bound multiples, as present in the most complex of living forms.

In the latter, many cells are joined to form organs in which highly specialized functions are enacted. Such functions work together in a coordinated manner in order to make the creature what it is. The more complex the creature, the greater is the capacity to adapt and to diversify in response to the environment. Some of these higher forms of living matter are composed of trillions of cells. One of the most important organs, the brain, is composed of billions of cells called neurons.

All cells possess the same general functions. They require nourishment which they must absorb or ingest. Within the cell, such nutritional elements must be digested and metabolized, to provide energy for maintenance and growth. Cells must be capable of excreting their waste products. In addition, all cells, except red blood cells and others without nuclei, have within their nuclei marvelous mechanisms for reproduction of their kinds. Unlike other cells, the neuron, after having reached its maximum embryonic growth, if destroyed, cannot be replaced. (Damage to the cellular components of the central nervous system becomes permanent.) The cell's propagating mechanism is guided and directed by DNA in such a manner that each generation of new cells has within it a proper assembly of chromosomes and genes to ensure the proper hereditary qualities of the cell. It has, in addition, all of the other components which are necessary for the cell's total functioning. These commonalities alone seem to be sufficient to establish beyond a reasonable doubt that the simplest and the most complex of all living structures are a part of one and the same combination of chemistry, physics, and energy. This entire area of interest is commonly referred to as molecular biology.

Is there other evidence that humans are a part of these evolutionary patterns? And what about the Darwinian hypothesis? What proof is there that bacteria, algae, protozoa, sponges and crinoids, crustaceans, fishes, amphibians, reptiles, dinosaurs, birds, and mammals – including primates and man – are all part of the same evolutionary progression?

The relationship of apes to the evolution of man is the area of evolutionary progression which caused the greatest controversy. What is there to sustain a belief in the ideas put forth by Darwin? And what is there to make acceptance of them tenable? The commonality of the cell as the basic unit of all living matter has already gained acceptance. Radiometric dating techniques and their applications to fossil structures have shown that there has been a succession of more and more complex forms, each bearing some semblance to its forebears. Fossilized remains have provided a tremendous resource, especially when they have been accurately dated. Studies in comparative anatomy have revealed related similarities and their variations.

What are some of the differences between man and primates? First, there is the matter of locomotion. The other primates do not stand erect they are quadrupeds. They move about by utilizing all four extremities. Humans stand erect and use only two lower extremities for locomotion, thus leaving the hands free.

Then, there is the matter of intelligence. Other primates are relatively limited in this respect. They cannot perform with the same degree of reason, logic, precision, and skills as do humans. They cannot design, make, and use complex tools because neither their intelligence nor their upper extremities' structure is as yet adapted to these purposes. Many of the primates are tree dwellers, and their dentition provides for the chewing and mastication of tree parts.

Well preserved fossilized remains have been discovered in diverse areas of the world: in India, Pakistan, Turkey and Iraq, Israel, western Europe, Java, China, East and South Africa, and elsewhere. The fossils resulted from remains which were deposited and preserved in sediment, mud, clay, or volcanic dust. The deposits provide identifiable geological strata which can be dated with accuracy. Often, sediments were separated by layers of volcanic ash which gave an additional dating mechanism. The technique of using the radioactive decay of potassium into its isotope of argon, was discovered by Willard Libby shortly following World War II. This proved to be one of the most important factors in establishing the evolutionary interrelationships.

The entire Leakey family (including the late Lewis, his wife Mary, and their sons Richard, Jonathan, and Phil), working mostly at the Lake Turkana area in Kenya, at Olduvai Gorge in Tanzania, halfway between the Nile and Lake Victoria, and in nearby Lake Laetoli, have been foremost in many important discoveries. Richard Leakey found, at Lake Turkana, a deep depression due to erosion which had been caused by the rushing river waters during the rainy season. The various deposits of sediment were separated by layers of volcanic ash. This area was made to order, for it not only provided certain important fossil finds but also gave a time picture of when skeletal remains had been a part of a living creature.

In the Olduvai Gorge stratified sediments similar to those found near Lake Turkana, several fossilized skeletal forms were uncovered by the Leakey team. Some of the fossils were similar to that of *robustus*, which we will later discuss.

We shall try to place the fossilized findings in proper perspective as they might fit into evolutionary progression. First, we must return to the chimpanzee, one of the most intelligent primates, the creature considered to be more closely related to man than any of the other anthropoid apes. The brains of chimpanzees are certainly more advanced than those of other primates. One other convincing bit of evidence has established this relationship. Alan Wilson and Mary King, working with the molecular structure of the blood protein, have discovered that there is over a 99% agreement between the protein elements of man and those of the chimpanzee. This relationship has not yet been established for any of the other primates.

Some of the primates wandered away from Africa; thus, the orange-colored orangutan is found on the Indonesian islands of Sumatra and Borneo. But, fossilized skeletons of the chimpanzee have never been discovered elsewhere than in Ethiopia, Kenya, and Tanzania, so these areas have been designated as the cradle of human origin. To establish the trends toward human evolution, let us go back prior to the time of the chimpanzee.

In the year 1961, Elwin Simons, a Yale paleontologist, while excavating in the once heavily-forested Fayum area of North Africa, discovered an almost complete skull and several other skeletal fragments. Radiometric techniques indicated that this creature lived 28 million years ago. It was given the name *Aegyptopithecus*, meaning Egyptian ape. Simons and others considered this to be one of the earliest of the primate apes. While working on the island of Lusinga in Lake Victoria, Lewis and Mary Leakey uncovered an almost complete fossilized skeleton. The creature to whom this skeleton belonged lived 20 million years ago, according to potassium-argon dating. After careful study by the Leakeys and others, it was concluded that this creature was intermediate between *Aegyptopithecus* and the primate apes. This and related finds are called the *dryopithecines*. These are the only two undisputed fossil findings of the pre-ape lineage. Others have been proposed but disputed, such as *pondaungia* and *amphipithecus* in Southeast Asia.

124 The period about one to two million years ago is of great importance because of the discovery of fossils representative of three distinct creatures of that time which advanced far beyond the ape-like stock, but were not yet human. In 1924, Raymond Dart, Professor of Paleontology at Witwatersrand University at Johannesburg, South Africa, made an important discovery while exploring a limestone quarry about 70 miles north of Kimberly. Here workers excavated a skull and a few skeletal parts which were imbedded in the limestone. The skull was that of a five-year-old primate, but it was somewhat more human than ape. The foramen magnum, the hole through which the spinal cord enters the base of the skull, was farther forward than that of any previous find. This indicated that in, all probability, the creature was capable of walking in an upright posture. The size and shape of the teeth were also more human than ape-like. Dart named this find *Australopithecus africanus*.

In 1936, Robert Broom, while excavating in the Transvaal area, found a fossilized skeleton similar to that which had been found by Dart. This creature was about 3'2" tall and was capable of an upright stance. The dentition was about the same as that of Raymond Dart's find. These two creatures apparently belonged to the same species and were, therefore, given the same designation.

In a nearby area, Broom discovered a much larger creature which weighed about 150 pounds. The skull of this creature was more massive, and the molars were much larger than those of *Australopithecus africanus*. This was the first evidence that the molars were capable of a crushing action. Otherwise, this was not much different from the two previous finds. But, because of its size, it was designated *Australopithecus robustus*.

Eventually early primates developed into man. The first adaptation was the erect posture which was found among all members of the triad hominids; the tilt of the pelvis, and the structure of the upper thigh. Then, there was the forward movement of the hole in the base of the skull through which the spinal cord enters. These features indicated that these creatures walked upright on two legs. The upper extremities were therefore free to develop into arms and hands for touching, feeling, and manipulating objects.

In 1961, Jonathan Leakey, another son of Lewis and Mary Leakey, discovered, in the Olduvai area, a fossilized skull and fragmented skeletal parts. When the bones were cleaned and assembled, Jonathan Leakey concluded that these findings fitted into the patterns of the *Australopithecines* except that the skull was a little larger and that the teeth were designed for an omnivorous diet. The brain case volume of this skull was about 700 cubic centimeters, while the skull of other Australopithecines had a capacity of only about 500 cubic centimeters. You will recall that the human skull has a brain capacity of 1500 cubic centimeters. This skull was a little more dome-like than those of previous discoveries, indicating that the forebrain, that portion of the brain where intelligence, memory, and association lie, had probably begun to develop.

Jonathan Leakey concluded that the creature he had found was the maker of several tools found in association with the skeleton. Louis Leakey therefore named it *Homo Habilis*, which means the man who is handy.

Mary Leakey, in an article titled "Footprints in the Ashes of Time," published in the April 1979 issue of *National Geographic,* reported that she had been informed by natives that, in an area of Tanzania located about 30 miles south of Olduvai Gorge, they had found some footprints. The area was known as Laetoli, because of the special red lilies which grow there. Between 3.8 and 3.6 million years ago, a nearby volcano had scattered its ash over this area. Rains came and moistened these ash deposits. Two erect, bipedal creatures--*australophithecines*--walking in a northerly direction had left their footprints in the ash. It only remained for the sun and time to harden the ash mud to a cement-like material and to preserve the footprints for Mary Leakey's discovery in the 20th century. These footprints were indistinguishable from those of modern man. Radiometric measurements indicated that the ash at the top of the footprint had been deposited there 3.6 million years ago, while that directly underfoot had been laid down 3.8 million years ago. Measurements of the length of the feet indicated that one individual was smaller than the other. Since the length of the foot is about 15% of body height, the size of the footprints indicated that the taller of the two was 4'8" tall, while the smaller individual was about 4' tall. Even though it had been surmised from skeletal evidence that Australopithecus walked upright, this was the first direct, positive evidence that manlike creatures had taken themselves from here to there on two feet. (Throughout the age of the motion picture industry, it has been commonplace for actors and actresses to make impressions in wet cement of their hands or footprints, which at present are preserved mostly for the amusement of tourists. What might paleontologists 3.8 million years hence think when they come upon such hand prints and footprints?)

Twenty-two fossilized skeletal remains have been found in the Laetoli area. One was that of a five-year-old child. Radiometric measurements of the strata in which these fossils were found indicate that they are contemporary with the individuals who made footprints in the volcanic ash. Fossilized skeletal remains of contemporary animals have also been found in a nearby area. These animals included elephants black and white rhinos, hyenas, and smaller rodent-like creatures. Such findings have had a profound impact upon the understanding of man's part in this evolutionary progression.

Richard Leakey and his team, in 1972, worked on a find that was located on the eastern shore of Lake Turkana. One of the team members, Bernard Ingeneo, picked up a fragment of a skull from a dry river bed. Further searches revealed many other fragments, which were then turned over to paleontologist

Meave Leakey (Jane Goodall's mother), and to anatomist, Bernard Wood. When these were properly cleansed and assembled, the team found that this skull had belonged to a very advanced hominid. This skull became the famous "Number 1470." It had a brain case volume of about 800 cubic centimeters, one third larger than that of the ape stock. The creature lived almost 2 million years ago. Other skull fragments found in an adjacent area indicated that the creature to whom these fragments belonged had a skull capacity of 900 cubic centimeters. Radiometric measurements of the strata in which this later skull was found indicated that its owner had lived 1 million years ago.

From Olduvai, Laetoli, and Lake Turkana, we go to the Afar region of Ethiopia, to join Maurice Taieb, of France's National Center for Scientific Research, and Donald Carl Johansen, of Cleveland's Case Western Reserve University. These researchers found stone tools which dated back 2.6 million years. In 1974, "Lucy" came into the picture of our primate family. Taieb and Johansen came upon her fossilized skeletal remains in the same area in which the tools had been found. She had been laid down in sediment over 3 million years ago. She consisted of an almost complete fossilized skeleton and was about 3'2" tall. She had been capable of upright walking, and her brain case volume was about one third that of modern humans.

Photograph of the skeletal remains of one of the earliest hominids, the Australiopithicus afarensis *called "Lucy."*

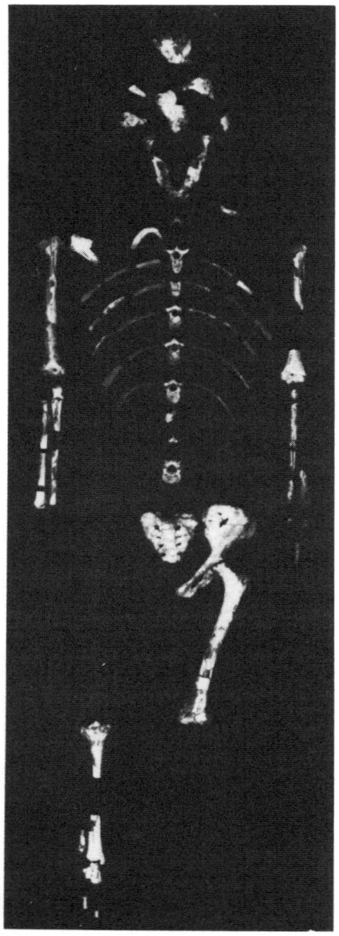

What do all of these findings mean with respect to the evolution of human beings? What can be garnered from these many discoveries of fossil remains, from their radiometric dating and from the areas in which they have been found? Well, they tell us, once again, that evolution has been a continuum, that it has progressed always from a simpler structural and functional status to one of further complexity, and that along the way some creatures have failed to adapt and have perished, while others have continued to produce progeny and have become part of that which is ever going forward.

The story is not complete. There are gaps, and hopefully at some future date they will be filled.

Part 4: HUMANS COME TO PLANET EARTH

Chapter 24: STONES AND BONES AND THEIR EVOLUTIONARY SIGNIFICANCE

There was a prophet who on many occasions, as he appeared before his followers, placed a grain of corn in the outer portion of his ear. When a dove flew down to pick out this choice morsel, it appeared to the faithful that the prophet was receiving messages from on high, with the dove acting as intermediary.

Our mythical travelers possess no such key with which they can receive immediate and revealing information concerning the future of evolutionary history. They find it necessary to tread the pathways patiently, and learn as they proceed.

It was not until about 2 million years ago, according to radiometric datings, that tools of stone were made. These tools belonged to creatures who were structurally advanced toward human emergence.

Some very primitive stone tools were discovered in areas adjacent to *robustus* skeletal finds. It was first thought that *robustus* made and used stone tools, but it was then learned that it was another creature who had came a little later and was somewhat more advanced who had made and used these tools.

Such early tools were discovered in Olduvai Gorge in Tanzania, on the eastern shores of Lake Turkana in Kenya, and in the Afar region of Ethiopia. These primitive tools had been broken from larger pieces of flint, basalt, and quartz. They were designed primarily for crushing, cutting, and scraping. Such tools, although not manifesting great precision and skill, required some degree of intelligence for the finding of proper materials, for the design and making of the tools, and for formulating the purposes and uses for which they had been intended. The first stone tool may well have occurred quite accidentally by the simple crushing of one stone against another, and once accomplished, such fragments could then have served the purpose. Some materials served better than others, and there is evidence that tools were made in such places where these materials were more abundant. Initially, tools may have resulted from a playful act. But the utilization for crushing, cutting, slicing, and scraping could not have been accidental and unintentional.

One other important piece of evidence cannot be overlooked. In several sites where these early tools were discovered, there were also found fossilized skeletal remains of animals which apparently had been scavenged and processed by means of such tools, and consumed. Even the long bones had been crushed, perhaps to obtain the marrow for food purposes. Included among these animal remains were bones of elephants, the white and the black rhinos, hyenas, baboons, and smaller animals such as hares and certain rodents.

Let us begin then to correlate some of these factors which seem pertinent to biological evolution as we approach the emergence of man. It was between 3 and 3.5 million years ago that *Australopithecus afarensis* appeared upon the scene. The skulls of these creatures had increased in brain capacity by about 400 cubic centimeters. But more importantly, as has been learned from making latex and plaster casts of the interior of such skulls, it is apparent that the increased skull capacity was due to an enlargement in the frontal portion of the interior of the skull (that is, in that area which contains the forebrain).

A simple increase in brain case volume cannot be totally correlated with intelligence. But, the area in which this increased brain capacity potential has occurred, the frontal portion of the skull, is very meaningful, and does assume importance with respect to intelligence. While the midbrain and the hindbrain are concerned with involuntary responses to stimuli and with regulating the functions of the vital organs, the forebrain contains the center for memory and for recall. Without the forebrain, thinking processes could never have come into being.

Many books could be written upon this one facet of evolutionary development, which started with that little marine-bound creature known as Amphioxus and with that tiny darkened spot on its forward end, the first evidence of a central nervous system. It took about 1.4 billion years for that little black spot to evolve into the central nervous system of man, which contains millions of neurons. Thus it was with these latter-day creatures: more would be required than a simple response to a stimulus, more than sitting on a tree branch to satisfy the urges of hunger by reaching for a fruit, more than hiding in fear from the flash of lightning or the violent noise of thunder or the nearness of a predator.

Indeed life was becoming more complicated, and abilities beyond those just described became increasingly necessary. Memory, thought, planning, motivation, and execution would play important roles in evolutionary progression before man, with the most magnificent brain of all living creatures, could emerge.

Homo habilis, over 2 million years ago, was already on the way to achieving this goal. The skull, with its increasing capacity and its modified contours in the area which would by occupied by the forebrain, had already demonstrated that this aspect of the evolutionary sequence was already well underway. *Homo habilis* could think: stone tools give mute evidence of that. *Homo habilis* could use tools, search for the best materials for them, and design, plan, and manufacture them. They were capable of gathering food and of processing it for consumption and nourishment. Tools were necessary for these purposes, and were made. These factors of intelligence played increasing roles among a succession of creature types prior to the advent of man.

During the last three million years of Earth's existence, when stone was the source of all tool making, there was a gradual increase not only in the variety of stone tools, but in the precision and skill with which they were made. By the time the first Cro-Magnon man came upon the scene 40,000 years ago, stone toolmaking had progressed from the primitive, simple, utilitarian forms to those made with precision and skill, to the expression of the individual tool maker's deepest thoughts in delicate and exquisite art forms.

By the time this latter occurred, the creatures involved had developed a complete brain structure indistinguishable from that of modern man. Indeed, Cro-Magnon was man. Stones and bones, the area in which they had been found, the association with the fossil remains of the creatures who made them, and the radiometric dating of such artifacts, all provide a time-related picture which establishes a pattern of evolutionary progression from creatures with somewhat limited intelligence to those with brain structures capable of performing the most complicated functions: namely, those we ordinarily associate with human beings.

Advance was gradual for the Australopithecines. *Homo habilis* and *Homo erectus* possessed brains of sufficient sizes and of the necessary component parts to meet their somewhat limited demands. Had the status quo remained, it is entirely possible that no further change in the size and structure of the forebrain would have occurred. However, evolutionary progress required that the development of the central nervous system of higher creatures keep pace with their environmental necessities. More and more pathways had to be established within the forebrain part of the central nervous system. *Homo habilis*, with a small forebrain, used tools for processing flesh for food. This little world of activity must have been quite simple indeed. However, as the horizons increased through the creatures' wanderings, they faced many new environmental situations. New accommodations had to be made to meet the rigors of climate, changing food sources, and other factors essential to their survival.

As the emergence of humans came closer and closer, almost the entire planet Earth had become habitable – but only because creatures were able to adapt and to diversify. Such has been the odyssey of the evolutionary development of intelligence and of the structures which have made it possible.

Part 4: HUMANS COME TO PLANET EARTH

Chapter 25: THE PRE-DAWN OF HUMAN EMERGENCE

As our mythical travelers approach the epoch from about 600,000 years ago to about 35,000 years ago (the end of Earth's glaciation) they have become acutely aware that the evolutionary processes, especially those related to the eventual emergence of humans, are moving at a much more rapid pace than had occurred in previous time spans. During those approximately 1.5 million years, there came into being several types of creatures which were more closely akin to humans than any which had previously existed. Not only did the types and numbers of creatures increase, but there is sufficient fossil evidence to show that these creatures dispersed themselves over widely separated areas of the Earth's surface.

Consider the monotony of existence which might have been present for early humans living in one area. Imagine the feeling of wanderlust, the desire to experience what might lie beyond the immediate habitat. It is not too difficult to assume that groups wandered away. Some individuals in these groups continued to distant places, while others, perhaps disillusioned by the hardships of travel, separated and set up places along the way. Perhap some pursued their journeys in entirely different directions. However it happened, some reached such faraway places as India, Indonesia, and China. They must have encountered new conditions such as differences in climate, and new and unusual sources of food to which they had to adapt. But it must be remembered that, by the time these precursors of man came upon the scene, certain capabilities had already become a part of their heritage. They could make and use tools. In some instances, they knew fire and employed it for several purposes. They could hunt and gather. They participated in group living. Some had constructed houses. This, indeed, was the pre-dawn of human emergence.

To demonstrate, retreat with our mythical travelers backward in time to 600,000 years ago, to a place on the island of Java, in Indonesia.

It was here, along the banks of the Solo River, that creatures lived and died who were more nearly akin to humans than any which had preceded them. Eleven fairly intact skulls and portions of two shin bones and two thigh bones were covered with sediment, became fossilized and preserved, awaiting their discovery by Eugene Dubois in the year 1891. The shin and thigh bones were not unlike those of modern humans, indicating that these creatures had been able to assume an erect posture and possibly could travel by means of bipedal locomotion. The skulls, however, were much smaller than those of modern man. The forward part of the skull was thicker, and there were prominent ridges above the eye sockets from which the forward part of the skull sloped sharply backward. The occiput, the back portion of the skull which housed the hindbrain, was pronounced. The position of the foramen magnum, the hole through which the spinal cord enters the brain, indicated that these creatures stood erect, corroborating the evidence of the leg bones. The angle at which the temporal bone and the jaw came together was similar to that found in the skull of the gorilla.

It was concluded that this fossil, called Java Man or Solo Man (later known as *Pithecanthropus*, and then in 1934 renamed *Homo erectus*) constituted a most important find, in that it seemed to be intermediate between *Homo habilis* and the much later Neanderthal Man.

Another important discovery was made in southern China. This was reported in the September 1930 issue of *Scientific American*. Near the village of Chou Kuotien, another individual lived and died about 500,000 years ago. His skull had also been covered with sediment, but nature had another method of preserving it. The creeping moisture around the skull contained calcium salts which, when deposited to create a block of travertine limestone, had imprisoned the skull within it so that it had become a part of the stone itself. It was discovered by W. C. Pei and Davidson Black who worked meticulously for over a year to free this almost intact skull. After evaluation by Black and other paleontologists, it was concluded that Peking Man, as this creature came to be known, was another important link in our understanding of the evolutionary trend toward man. The skull was larger and had humanlike characteristics far more advanced than the skulls of Java Man. In an area close by Chou Kuotien, several

From left to right a female gorilla, a female Sinanthropus pekinensis, *and a modern male.*

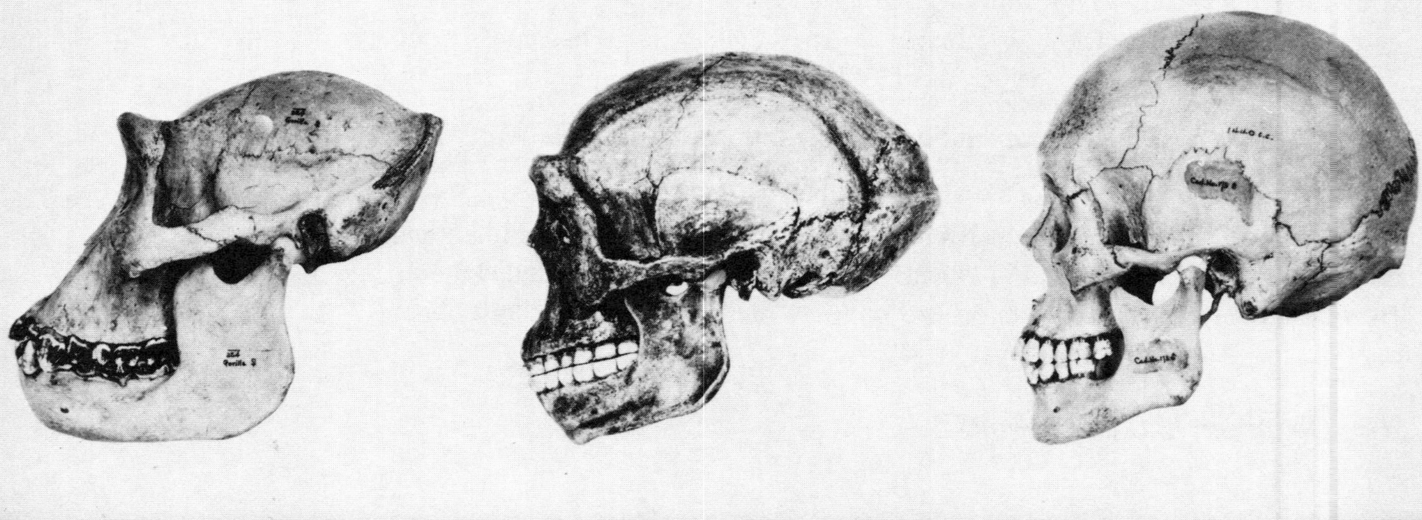

other skulls were excavated. These skulls were similar to that of Peking Man, with one almost unbelievable manmade modification: the hole in the base of the skull, the foramen magnum through which the spinal cord had passed had been carefully and skillfully enlarged, indicating that without question other creatures had performed this enlargement by a purposeful act. In all probability, this was a means by which the inner content of the skull, the brain, could be removed.

How can a discovery such as this be interpreted? Could these creatures who lived half a million years ago been up to some ritualistic procedure? Could they have removed and consumed the brains in order to establish a continuity with their immediate ancestry? Is this the earliest evidence of organ eating as has been practiced in much later aboriginal cultures? Since there are no means by which these questions can be answered conclusively, one can only resort to conjecture. But, no stone tools by means of which the enlargements could have been made have ever been found in the vicinity.

Now, we shall go half way around the world, to southern France, to a place called Terra Amata, overlooking the shores of the Mediterranean. There is a hillside which is now incorporated into the suburban area of Nice. On this hillside 400,000 years ago creatures with the most amazing capabilities resided. On this site, with the beauty of the blue Mediterranean water always before them, they built a dwelling. It, too, was covered over and preserved, and had it not been for some workmen who were excavating the site for the construction of a luxury apartment complex, this story could never have been told. They found that holes had been carefully dug into the earth and wooden poles had been inserted into them. Around the base of each pole, large stones had been carefully piled to give support. These poles had formed a basic wall structure, and it is not difficult to imagine that animal skins or other perishable materials were draped from pole to pole in order to keep out the cold, the wind, and the rain. Inside the structure, a hearth had been constructed on one side. In front of the hearth no bone or stone fragments were found, nor any other debris. Across from the hearth, on the opposite side, there were fossilized skeletal remains of elephant, deer, wild boar, and smaller animals. Near to these bones were found stone tools for crushing, cutting, and scraping. Unfortunately, there have never been found any fossilized skeletal remains of the creatures who built this habitat and who dwelt there. But the evidence for creature involvement in such advanced technology is everywhere apparent within this structure. Even the site must have been carefully chosen.

In the mountainous region of Iraq, near the Turkish border, at a place which has the intriguing name of Shanidar, there is a cave where the first evidence of ritualistic burial was found. The fossilized skeleton of a Neanderthal was found, lying on his side. A litter had been carefully woven using the small branches of the woody horsetail, a plant indigenous to that area. Around where the corpse lay, many identifiable pollen grains of well known flowers were discovered. All of this indicated that the deceased was placed there with a sense of loss, perhaps of grief, and most certainly with love and affection. This all happened 60,000 years ago.

Now we must travel back to France where 50,000 years ago advanced creatures lived and died. Here, at La Chapelle aux Saints, railroad workmen uncovered a burial site. A Neanderthal had been placed in a grave built of stones cut and shaped by hard flint tools. A haunch of meat had been placed in the grave, together with several flint tools. This burial site reveals to us many things. As at Shanidar, there was a ritualistic internment. However, it appears that there was something more; perhaps the earliest evidence that creatures who had lived 50,000 years ago had a sense of the hereafter, as provision was made for both food and tools to accompany the individual on his or her journey into that realm beyond life on Earth.

These findings help us in the development of a concept as to when our evolutionary ancestors developed thoughtful planning and execution, as opposed to reactions only in response to stimuli. How closely are these discoveries related to intelligence and to the higher emotions? Did the first evidence of such become manifest with *Homo habilis*, the stone tool maker? Was the common good a motivation for group hunting, gathering, and living? Were all of these manifestations dependent upon increasing cranial capacity, particularly in the area of the forebrain? It is difficult to conjure up answers to all of these questions, simply because the evidence is too scant. Bones and stones, tools and burial sites, cannot and do not tell all. However, the building of a habitat such as that which was constructed on the hillside near Nice was something more than random assembly of available materials.

How about the intangibles, the manifestations of grief, tenderness, love, affection, and a sense of the hereafter? These qualities are difficult to determine by means of archaeological artifacts. However, both Shanidar and La Chapelle aux Saints cannot help but leave us with the impression that some aspects of intelligence and human emotion had already begun.

A short journey from France to Western Germany brings us to the Neander River valley near Dusseldorf. In the year 1856, some workmen were quarrying limestone about 30 meters above the river bed. In the process of their work, they broke through the walls of a cave. Here they found the fossilized skeletal remains of a humanlike creature. Not fully realizing the evolutionary value of their find, they were somewhat careless, and all that could be salvaged was portions of the skull and the jaw, and a few other fragmented remnants. After cleansing and assembling, these remnants were studied and evaluated by S. Meyer of Bonn, Germany, and other paleontologists. When they were compared with Gibralter Man, another Neanderthal type, it was concluded that these two creatures were very special, in that their fossilized remains provided evidence of creatures more nearly related to man than any which had been previously found.

What were the characteristics which determined this special significance? First of all, Neanderthal Man (the name means Neander River Valley Man) and Gibralter Man had large skulls, with volumes between 1600 and 1700 cubic centimeters. (By comparison, modern man has an average skull volume of 1400 cubic centimeters.) These were the largest skulls yet found. As in earlier forms, there were very prominent ridges above the eye sockets, and between the ridges where the skull sloped somewhat backward there was a depression. The same angle where the jaw and the temporal bone occlude, as had already been found in Java and Peking Men, was present here also. The face was flat and broadened, while the jaw was smaller and did not protrude as far forward. The foramen magnum was set far forward, indication that these creatures had assumed an erect posture. The dentition was identical with that of modern man, except for the molars, which were a little larger and which had a little greater pulp capacity. The arms, apparently, were longer.

Over 100 fossilized skeletal remains, many of them quite complete, have been discovered in sites ranging over a wide territory, from Pakistan and the Middle East, Israel, the Balkan states, Italy, France, Spain, and central Europe. These creatures which, except for minor variations, resembled Gibraltor Man and Neanderthal Man, all lived within this same time span (from 150,000, to 36,000 B.C.).Great excitement reigned for over 40 years because of the so-called Piltdown Man, which was considered to be even more advanced than *Homo neanderthalensis*. The skull was not unlike a human skull, while the jaw was like that of the primate apes. Eventually, it was found that a prankster had put together a human skull and the jaw of an ape to accomplish this forgery. By the time it was discovered to be a forgery, however, much character damage had been done.

When these several advanced characteristics are taken together and correlated to the time span during which the creatures possessed them, it can be concluded that *Homo sapiens neanderthalensis* was more humanlike than any creature which had existed before. One thing is certain in our review of evolutionary history. There is no evidence whatever that any of the creatures that lived up to 35,000 years ago were the direct ancestors of humans. There is ample evidence, however, that structural changes which caused creatures to become more and more humanlike, when correlated with the progression of time, provided the background for the concept of branching which could then lead to the emergence of man. However, the links are missing between the last of the Neanderthal people and the beginning of the epoch during which the magnificent Cro-Magnon people prevailed. These were the direct ancestors of human

136 beings. There must have been a connection between *Homo sapiens neanderthalensis* and Cro-Magnon people, but as yet no evidence of such has been discovered. It is reasonable to assume that there must be such links with our ancestral past. We can hope that the fossilized remains of a creature will be found to substantiate the hypothesis that humans are the product of an evolutionary process.

Part 4: HUMANS COME TO PLANET EARTH

Chapter 26: THE MAGNIFICENT CRO-MAGNON MAN

Along the pathways of cosmic time, our mythical travelers have had many doorways to discovery unlocked before them. Behind each portal there has been present some new aspect, some new detail, and some new scientific truth which gives substantiation to the theory that humans, together with the primate apes and the various intermediate forms, evolved from a common stock; that *Homo sylvestris*, or the Forest People (as the apes were sometimes called) continued to produce progeny like unto themselves even into this, our 20th century.

However, somewhere in our ancestral past, modification of basic structure, form, and function has occurred. Modern genetics tells us that these changes could have been brought about by adaptation or diversification due to environmental changes, or because of beneficial or disadvantageous mutations which then became imprinted in the genetic code of DNA and were passed on to progeny.

This is the principle which Charles Darwin and others, such as Edward Blyth, Alfred Wallace, Sir Charles Lyle, Thomas Huxley, and even the grandfather of Charles Darwin, Erasmus Darwin, found and formulated. Thus, as we approach the emergence of human beings, certain evidence is there to guide us in our thinking. There are the tools of stone and bone, and later of tusks and antlers. There are the fossilized bones of animals which man-like creatures consumed. In some cases there are shelters and evidence of fire. There are radiometric techniques of dating these artifacts, so that a time phase can be attributed to them. And lastly, studies in comparative anatomy have augmented our knowledge of evolutionary biological progression.

An epoch which began 35,000 years ago produced creatures which were not only somewhat similar to but were indeed exactly like humans. These were the magnificent Cro-Magnon people.

In the plains of Les Eysies in the Dordogne region of France, in 1868, some road-building workmen, removing soil beneath an overhanging limestone cliff, discovered the fossilized skeletal remains of four adults and an infant. The workmen notified scientists, who were able to remove the almost complete skeletons. One was an adult male appproximately 35 years of age and about 5' 8" tall. In addition, there were two adolescent males, and a young female whose two- to three-week old infant had been buried along with her. These skeletal remains were in no way different from those of modern humans. In addition to the skeletons there were found skillfully made flint tools. There were also shells and animal teeth which had been perforated with small holes, in all probability so that they could be worn as pendants, necklaces, or bracelets for personal adornment.

A legend concerning this site in which the bodies were discovered relates that a hermit named Magnon had lived there. Thus, the designation of "Cro-Magnon" was given to these fossilized remains and would subsequently include all other peoples until the beginning of the age of metals.

At last, after 4.6 billion years of Earth's existence, the first humans had arrived. During these last 35,000 years further change would be concerned with purely social and cultural progressions rather than with physical or biological progress.

Many paleontologists are of the opinion that the last of the Neanderthaloids, such as those represented by the fossil remains found at Mladec in Israel, were a little more humanlike than their forebears and could almost be considered as transitional forms between Neanderthal and Cro-Magnon. Certainly their tools were a little better made, and some of them were made of bone.

But what about these Cro-Magnon people? What about their capabilities, which no other creatures before them possessed? In Czechoslovakia, a 27,000 year old site now known as Dolni Vestonice was discovered and excavated in 1951. Here, a remarkable people built a village. Nearby was a running stream from which they could obtain water. There were five buildings surrounded by a protective wall which had been constructed by placing pointed nine-foot long mammoth tusks of ivory at intervals, just as a modern-day farmer might space his fence posts. Bones of tremendously large animals were placed between, together with rushes and soil, to complete the wall.

Four of the five buildings, in all probability, were living quarters. These four faced each other, and in the central space between them a huge fire had been kept going, possibly to frighten off marauding animals. The basic structural characteristics of these buildings were in many ways similar to those which were found earlier on the hillside overlooking the Mediterranean in the suburb of Nice, France. Upright poles were placed in holes in the earth and supported by large stones and bones. They tilted toward the center of the top. Animal skins had apparently been sewn together and stretched over the outer surface of these supporting uprights and weighted down along the bottom edges by stones and bones. The largest of these huts measured 50' x 20'.

Holes were dug into the earthen floor to accommodate five hearths, so the entire structure could be warmed. One of these hearths was equipped with two large, long bones such as might have supported a spit for the roasting of meat. Many well-made flint tools were found in the area.

One fantastic find was that of a hollow bone into which three holes had been drilled. This was indeed a three-note whistle, somewhat after the fashion of the latter-day penny whistle. This is the first evidence of anything resembling a musical instrument.

The fifth building was excavated into a cliff at the top of the hillside. The entrance, facing the other huts downhill, was made of stones piled one upon the other. Here was also a hearth, but one of an entirely different kind. Atop this hearth there was a dome-shaped earthen cover. For what purpose had this special kind of hearth been constructed? The answer soon becomes apparent, for on the floor there were numerous pieces of pinched off bits of clay: this was a ceramic kiln. Some pieces of clay still show the thumb print of a ceramicist who worked here 27,000 years ago. In addition, there were numerous broken pieces of animal and human figurines which had been fired. Fortunately, some figurines have remained intact. There was a lion in which a hole had been made directly over the location of the heart. Could such a figurine as this have been used to show where the most effective wound should be made during the hunt? Possibly so. Some female figurines, which have come to be known as Venuses, had large buttocks and thighs and extremely large breasts, such as have been found in the fertility figurines of other cultures, particularly those of Latin America. In this case, the lower extremities were pointed. They could have thus been stuck into the earth to assume an upright position where they could serve as symbols of fertility or of protection.

One of the most remarkable discoveries in this ceramic workshop at Dolni Vestonice was that the ceramicist had learned to mix clay and powdered bone. In this manner a higher and more even temperature could be attained in the firing process, and the quality of the product would be superior to that made with clay alone. 15,000 years later the Japanese and Chinese discovered this process for making bone china, and it was 20,000 years hence before the Mesopotamians discovered that their history and their records could be permanently preserved by inscribing and baking tablets of clay.

Cro-Magnon people sought for and discovered the best sources of flint for tool making, and they developed the fine art of knapping. In this process, a core of flint a few inches in diameter and from two to ten inches in length was taken from its source. Then, by applying proper pressure or a tap to the end of the core, flakes of the entire core length would fall away. Skilled knappers could make precision tools much to their liking. The famous laurel leaf design, a ten-inch blade of delicate thinness and beauty, is a good example of the artistic element among these Cro-Magnon tool makers. It was a manifestation of the highest form of knapping and may have played a role in some ceremonial or ritualistic procedure. Several of these were discovered at diverse sites.

Cro-Magnon people also used other materials for tools and for works of art. Bones as well as antlers were fashioned into handles for various implements, thereby extending the effectiveness of such tools. Nine-foot long tusks of ivory of the mammoth were extensively employed. New tools came into existence in order to work these materials. A fantastic little instrument known as the burin was a tool with a wedge-shaped chisel end. Present-day lithographers use such a tool for inscribing copper or stone in making their prints, and glaziers employ it for cutting glass. With the burin, Cro-Magnon people were able to inscribe a circle around a piece of ivory or a bone and then break the piece precisely into the parts desired.

A laurel-leaf design flint blade, made by a Cro-Magnon tool maker.

Needles could be made from ivory and bone by repeatedly drawing two parallel lines with the burin, each going deeper and deeper until the slender fragment could be lifted away. With further polishing and pointing, and with a hole drilled in one end, the finished product could be used for sewing. These people, exposed to the cold, had to clothe their bodies in animal skins. Doll-like figurines recovered from the steppes of Russia revealed that these people dressed much as do our 20th century Eskimos, even to the parka and the boot. With a combination of intelligence and tools, these people were able to build better, larger places of habitat for both security and comfort. They learned to make and use fire in ways which their predecessors had not. In Belgium there were found small pieces of iron pyrite which had been worn by repeated striking with a piece of flint. Flint upon flint produces an ineffective spark; however, flint and iron pyrite make a spark of sufficient intensity to ignite tinder. This enabled people to carry their fire making tools with them.

Cro-Magnon people also learned how to construct more efficient hearths, so the flame would produce greater heat. They could not have known that oxygen was the element necessary for combustion. Nevertheless, they utilized this principle, for beneath the hearth where the coals would have been, they dug parallel depressions with a larger fanlike trough leading into them. A continuous flow of air could then circulate up between the burning coals. It is thought that, in some areas where wood was scarce, bones had been used as fuel.

Hunting and gathering had by this time become necessary for survival. This is attested to by the huge number of animal bones found near excavated sites and by the numerous pollen grains within the shelters.

Cro-Magnon people possessed the same intelligence potential as that of modern 20th century humans. They learned to respect nature, to foster some control over its adverse elements, and to use it to their advantage where possible.

Each new discovery provided for them a degree of social and cultural advance. They learned the beginnings of architecture; the design and planning, choice and assembly of materials, and execution of building projects upon carefully chosen sites. They learned how to make tools, skillfully and precisely, out of the best materials then available. They learned not only to make fire but to use it in the most efficient way. They developed a musical instrument – the three-hole bone whistle – and the soul of art was incorporated into toolmaking and into decorated bone and ivory.

Paintings on the walls of caves were perhaps their ultimate art form. Here were depicted scenes of the hunt. The irregularities of the walls were used to express the body contours of the animals painted upon them, and the pigments were natural substances such as yellow ochre and common metallic oxides. The caves in France, and at Altamira, near Santandar in the northern part of Spain, are excellent examples of the artistic capability of Cro-Magnon man.

As more and more people grouped together in common living sites, hunting and gathering had to be done on a more efficient scale, especially as they wandered into the more severe climatic ranges. With long winters, stores had to be of sufficient quantity and quality to maintain them. It might have been that as spring came and they witnessed the yearly emergence of plant life, the advantages of agriculture became apparent to them. No one knows just when planting and harvesting began, but it may have been in the Cro-Magnon time, about 10,000 years ago.

Throughout their encounters with evolutionary progress, our mythical travelers witnessed many aspects of structural and functional modifications which were necessary for adaptation to environmental change. Our travelers are aware that a creature who possesses a large brain is not necessarily super-intelligent. The brain case of humans has an average volume of about 1500 cubic centimeters, but there are large variations which have no relationship to intellectual capability. Ivan Turgenev, the Nobel Prize winning Russian author, had a brain which measured slightly over 2000 cubic centimeters, while the brain of another famous writer, Anatole France, measured only 1000 cubic centimeters; yet no one would deny the intellectuality of either.

There are many other cases in which no relationship between brain size and capability can be established. Generally speaking, larger creatures have larger brains, and vice versa. But even this is not always so. Witness the dinosaur with its great mass and diminutive brain. The most intensive neurological

A cave painting of a deer hunt found in Castellon Spain.

research has established that it is the internal organization, and not the size of the brain which can be correlated with the degree and quality of its function.

Comparative anatomical and physiological studies of the brains of many species have made it quite clear that each species is equipped with a brain of sufficient size and internal organization to meet its own specific perspective and respond to its environment. Each progression to a new and more complicated environment has necessitated some addition to both the structure and the function of the central nervous system.

Simple responses to minimal environmental problems could be satisfied by brains somewhat less complicated than those required by more complex organisms. Each creature must react to the various facets of the world surrounding it in order to survive, grow, maintain itself, and reproduce. And the central nervous system, including the brain, must be geared to meet the specific demands of the members of each species.

As creatures were subjected to changing environmental patterns, even further modifications in central nervous system structure and function became imperative. Throughout the age of the mammals, at least, such progressive central nervous system changes did occur. Hundreds of fossil skulls from our ancestral past have been discovered, studied in detail, and time-related by radiometric means. This has shown that there has been progressive modification of the brain, especially in the part known as forebrain.

The social and cultural contributions of human's ancestors, as described earlier, had a tremendous impact upon the lives of all future humans. *Homo sapiens*, the wisest of the wise, had finally arrived with a brain so intricately and perfectly arranged that it permits intelligent performance far superior to that of any creature in the ancestral past.

Cro-Magnons could think, remember, and call forth from memory at will. These creatures could sort perceptions and integrate and correlate them. They could choose and plan and execute. They were capable of the highest forms of emotional involvement and artistry. Even more, they could recognize themselves as individuals who possessed all of these capabilities. Early people learned that group living was mutually advantageous. The digging stick and the carrying pouch were the tools of women for securing roots, seeds, nuts and fruits, while the men used their flint tools for killing and butchering animals. Such plant and animal foods were returned to the place of habitat, prepared, and distributed for the common good.

We can picture a scene which might have ensued following the hunting and gathering. The hearth has a flaming fire over which is being roasted a haunch of meat. People are gathered about, intrigued by the magic of the flame and enjoying its comfort and savoring the smell of the food which is about to satisfy their hunger. Such peaceful scenes as these must have been quite routine among the gathering and hunting peoples.

However, there must have come a time when human intelligence perceived that hunting and gathering in locations remote from their shelters was frequently inconvenient. They had seen seeds sprout in the spring and produce new plants which in turn produced seeds for harvest. Such seeds could be gathered and preserved. When spring came, these seeds could be planted in places close to where they lived. Likewise, animals could be corralled and permitted to breed; thus both plants and animals could be readily available.

Artist's conception of Cro-Magnon man, Homo sapiens sapiens.

On the opposite page is an artists's rendition of a cave painting found in the Cueva de los Caballos (Cave of the Horses).

Part 4: HUMANS COME TO PLANET EARTH

Chapter 27: EARTH'S BOUNTIFUL METALS

Metals were first unearthed, smelted, and utilized about 8,000 to 9,000 B.C., at approximately the same time that prehistoric people were developing the techniques of agriculture.

You will recall that within the vastness of the universe, billions of suns were created. In the process, all of the known atomic elements (except plutonium, which is manmade) were created. Some of these elements grouped themselves into stony and metallic meteorites. Whirling in space, 93 million miles from the sun, the Earth had a gravitational attraction because of its mass. Stony and metallic meteorites in space were therefore attracted to its surface, and they accumulated around this flaming core. The metallic meteorites, because of their greater densities, came to lie nearest to the core, while the stony meteorites were superimposed. Volcanic action, earthquakes, and the rising of land masses from the sea brought these metals to the crust of the Earth, where they became available to humans. Copper was the earliest of these metals to be discovered.

Dr. Colin Renfrew, Professor of Archaeology at Southhampton University in England, had long theorized that the Bulgarians had been the first people to mine, smelt, cast and utilize copper, about 8,000 years ago.

In the Maritsa Valley in southern Bulgaria the landscape is dotted with burial mounds covering the remains of fallen soldiers. There are also much larger settlement mounds whose construction began about 6,000 years B.C. One of these, at the site named Kara Novo, is of particular significance with regard to the early discoveries of copper. At this settlement, houses were made of mud plastered over wood, and had to be replaced about once in each generation. About six inches of mud and debris were left beneath each new structure. Many such renewals left a mound approximately 50 feet in height. Professor Georgi Georgiev from the University of Sophia dug into the side of one settlement mound and removed a wedge-shaped section. Standing at the bottom, one could look up and see the chronological layering of each generation's debris. In the lower layers, there were flint and bone tools and pieces of pottery which were made by stone age people. In succeeding, higher levels, there were remnants of the copper age, including pins, beads, and a crucible which had contained molten copper. Excavations at nearby burial sites, which were constructed at a much later date, produced bronze axes of elaborate design. Carbon-14 testing and tree-ring dating have established that the copper and bronze age of Bulgaria was much earlier than that of Troy or any of the Aegean sites.

Three prehistoric copper mines have been recently discovered: one at Catal-Huyuk in Turkey, another at Ai-Bunar in Bulgaria, and a third at Rudna-Glava in Yugoslavia. These sites are remarkable in that the tools which the miners used to remove the pure copper from the stony surroundings were identical. There were teardrop-shaped stone axes or mallets like those that were fashioned by people many thousands of years before. They were used to crush away the rock from the veins of copper. Also, there were pointed antler tools for further removal of the copper from the crushed stone. Metallurgical procedures were not yet a part of the knowledge of these early peoples who lived about 9,000 years ago, for the only fabrications of copper consisted of beads and flat pieces which might have been used for bracelets and which could be made by simple hammering and shaping. It is Dr. Renfew's opinion that discovery, fabrication, and use of copper in what is now Bulgaria predated that of western Europe, and that the two arose independently.

Another area in which copper proved to be abundant was in Israel in the Arabah region at a site called Timnah in the Negev Desert. Here, green stones called malachite abound. The Old Testament Book of Deuteronomy (8:9) describes this as a place "where copper can be dug from the hills." These green stones did contain copper, but it could not be as easily retrieved as from the mines of Bulgaria, Yugoslavia, and Turkey. First, it had to be crushed to pellet-sized fragments, then placed in a bowl-shaped clay container and fired over a wood charcoal fire to reach a temperature of over 1500 F. This high heat could be attained only by playing a stream of air over the charcoal. There is evidence that this was accomplished by human lung power created by blowing through a ceramic-tipped tube. After firing, each pellet had to be broken to retrieve the small bead of copper which was contained within.

A thirty-five pound copper boulder, found in a glacial deposit in Wisconsin.

As time passed, the Egyptians learned of these rich copper deposits at Timnah and gradually took over the copper industry. They sent many men to extract malachite from the hillsides. To produce more charcoal, they ruthlessly cut down cedar trees. They built larger and better bowl-like receptacles of clay. Their fires were brought to even higher temperatures by means of hand-operated bellows made of animal hides. Layer upon layer of pellets were placed in the bowl as the firing progressed. When the black cinders of slag had been removed, the copper was in the bottom of the crucible.

It is estimated that by the year 1200 B.C. Egyptian workers were producing over 200 pounds of copper each day. Men soon learned that smelting was not all that was necessary to bring out the superb qualities of the metal. Copper from one area seemed to possess qualities somewhat different from that of another. It would seem likely that copper workers would have chosen regional coppers which, when smelted, would produce the particular texture or degree of hardness best suited for the intended purpose. They could not have known that such differences were due to impurities specific to the copper of each of these areas. In modern times it is no secret that two metals can be smelted together in order to produce a product of desired quality.

Arsenic was one of the common contaminates of copper. When smelted, it produced a much sought after degree of hardness and compactness in the bronze which resulted. However, arsenic is deadly and many of the smelting workers must have died of arsenic poisoning. Perhaps they thought that the wrath of the gods had been called forth to punish those who robbed the Earth of its treasures. But arsenic was not present in all regions, and it became evident that some coppers could be smelted without the deadly effect and still produce products of unique quality.

It is now known that tin and lead combined with copper produce excellent alloys. Lead contributes softness. Tin contributes softness or hardness, depending upon the percentage used in the smelting process; for example, hard bell metals contain 25 percent tin.

Long before the science of metallurgy came into existence, early people used metals to produce articles of superb quality for both utility and art. Consider the Cave of the Treasures, located in a high cliff overlooking the Dead Sea. The treasures remained undiscovered until recently probably because this cave was accessible only by means of a rope from above. Fortunately this cave, found in 1961 by archeologists from Hebrew University in Jerusalem, had escaped plundering by the Bedouin. On the floor of the cave there were bits of papyrus, parchment, and pottery with Hebrew words and inscriptions and fragments of glass, leather, and textiles. It is thought that this cave was occupied at the beginning of the Christian era by Jews attempting to escape persecution in Rome. At a much deeper level the archeologists found hearth stones, pottery, and supplies of wheat and barley. Radiometric datings established this occupation at about 3,000 years B.C., also when the Jews were fleeing for survival. But the most amazing discovery was at a deeper level still, where, after removing a slab of stone, a treasure almost beyond belief lay before the diggers.

A total of 429 articles had been carefully wrapped in straw matting and meticulously deposited there. All but 13 of these items were made of bronze, using the ancient technique of compounding copper and arsenic. There were ten crowns, 240 mace heads, several standards such as might have been used in rituals, and a variety of chisels and axes. The crowns, maces, and standards were, in all probability, items of religious significance, perhaps brought here from an endangered temple for safekeeping and preservation. Those who did so performed their services well. If the second level of discovery was dated at 3000 B.C., and the combining of copper and arsenic was practiced by the earliest smelters, then it could be that the amazing bronze artifacts found in the cave of the treasures could have been fabricated by peoples who lived as far back as 5,000 to 6,000 years B.C.

Bronze not only revolutionized the entire toolmaking industry but found its place as material for weapons of war, for magnificent works of art, for ceremonial and religious purposes, and to accompany rulers and their entourages on their journeys into the hereafter. It would no longer be stones, bones, tusks, and antlers but metals which would dominate the future. Other metals, such as iron, tin, zinc, lead, chromium, nickel, and aluminum would play their roles, especially following the industrial revolution.

Gold, silver, and platinum had unique qualities, too. First of all, they were rare. Second, they were soft and easily workable. In addition, gold glittered and silver and platinum glistened. They became much sought after as wealth, for the fabrication of articles for adornment and display, for the sacred vessels of temples and churches, and for the titular articles of royalty. The oldest treasury of gold artifacts was found in 1972 near the city of Barna in Bulgaria. Dr. Colin Renfrew describes this discovery in an article titled, "Ancient Bulgaria's Golden Treasures," which was published in the February 1980 issue of *National Geographic*. Except for a tractor operator who was digging a trench in which to bury an electric cable, these treasures might never have been discovered. His machine brought to the surface small thin sheets of what appeared to be bracelets of a glittery material. There were also bronze axes. It was later determined that he had dug into a very ancient cemetery. Radiometric dating established that these grave goods had been placed there between 4,200 and 4,600 years B.C. The more elaborate golden treasures of Troy which had been discovered by Heinrich Schliemann in the 1870's were fabricated 1500 years later than those found at Barna. By that time, the Thracians who had occupied these areas had become expert in goldsmithing.

The earliest fossilized human skull discovered in the New World was found recently in the area of Los Angeles. According to radiocarbon dating, the creature to whom it belonged lived 23,000 years ago. No one knows how early New World men discovered gold and silver. It was not known until the time of the Spanish conquistadors that the Peruvian, Columbian, and Mexican treasures even existed.

Your author had the privilege of viewing, examining, and photographing some of the outstanding collections and can attest to the superior artistic expertise of those who made them. One can think only with great sadness of how the early conquerors stole and smelted into ingots articles of magnificent beauty in order that the gold might by shipped more easily back to the rulers of their native lands. Of all the golden creations of our New World people, only a few pieces remain to be displayed in the museums of the world. The earliest discovery of these precious metals along with copper and bronze dates far back into prehistory, but they are still treasured today for the precious things they are.

The more utilitarian metals, all of which were to play an important part in man's future industry and technology, were not discovered and appreciated until much later – but that is another story. For the moment, our mythical travelers are overwhelmed by their encounters with prehistoric metalsmiths. They were the discoverers, the extricators, the earliest of the metallurgists who smelted, combined, and refined the metals, eventually fabricating articles of utility and beauty. They were prehistoric in the sense that they left no written records; yet their accomplishments provide us with plentiful evidence of how far they advanced both socially and culturally.

Part 4: HUMANS COME TO PLANET EARTH

Chapter 28: THE LANGUAGE OF HISTORY

Throughout time history has worn many hats. The little planet on which we live, its preparation for the advent of life, the fossilized remains of many of its numerous inhabitants, and later, the tools and works of art which advanced creatures have made all provide a record – not in terms of written language – but nevertheless substantial.

There must have come a time when creatures who were not yet human found it necessary to communicate with each other. For example, once again we can go to that hillside in southern France, overlooking the Mediterranean in the suburban area of Nice. There at Terra Amata creatures chose their site, and having found the materials, assembled and constructed rather complicated houses. Inside, they made a hearth where they could enjoy the magic and comfort of fire and consume food which had been roasted over it.

Some form of communication must have been necessary among those engaged in this project. Likewise, as far back as 80,000 years ago, the ritualistic burials at Shanidar in Iraq and at the La Chapelle aux Saints in France, both of which have previously been noted, must have required group effort which could only be accomplished through communication.

The magnificent cave drawings and paintings which were discovered in northern Spain at Altamira and at various places in France, and the ceramic miniatures of both human and animals which were created at Dolni Vestonice in Czechoslovakia over 27,000 years ago formed a means of visual communication. Valuable as all these records have been, they could never be considered to be commensurate with those that have been provided by the written language.

It was not until about 5,000 years ago that writing developed. The non-Semitic Sumerians of Mesopotamia devised cuneiform writing, a non-alphabetical system of pictograms, ideograms, phonograms, and syllable-signs. The Semitic Akkadians (that is, the Bayblonians and Assyrians) later adapted the Sumerian cuneiform script for their own language. The signs, or characters, are arrangements of wedge-shaped strokes impressed with a stylus into wax, clay, brick, or stone. There were thousands of signs originally, but they

Foundation cylinder of Nebuchadnezzar with inscription, clay, Babylonian. Lugar-Marad temple at Wanna-es-Sedum, ca. 600 B.C.

eventually diminished to about 800 or 900, for the simple reason that the same syllable could stand for increasing numbers of word parts, and one sign could be used as pictogram, ideogram, phonogram, and syllable. Cuneiform writing was in use from the middle of the third millennium B.C. to about 75 A.D.

The first cuneiform writings were made in wax poured on slabs of stone or wood. A few examples of this have been found at Uruk in the land of Ur. These remarkable people also learned that such impressions on tablets of soft clay, when dried and kilned, could provide a permanent record of what they had written. The stylus produced letter impressions with an angle on each side, causing one early archaeologist to remark that the writings appeared as though a bird had walked over the clay when it was soft.

Excavations in Sumer, Babylonia, and Assyria have produced thousands of these written-upon tablets of clay. Unfortunately, many of them were fragmented and broken. Some of the damage was done by not-too-careful early excavators, but most of the damage was done by tribesmen who plundered and robbed ancient cities, palaces, and temples.

A sufficient number of such tablets have been recovered, some intact and others capable of being pieced together, so that, when studied and deciphered, the tablets provide a history of how these people lived and in what they believed. On these baked tablets of clay they recorded legends, myths, a history of the reigns of their kings, epics, and poems, together with religious incantations, psalms, and hymns. There were also records of stores, supplies, and transactions. One interesting tablet contains the story of a father's instructions to his son as to how he could prosper in agriculture. Many other incidents, stories, and tales have also been deciphered.

Had it not been for the foresight of the Assyrian, King Assurbanipal, many of these tablets would not now be available. He directed his emissaries to go to the great cities of Uruk, Ninevah, Minrud, Babylon, and others, and collect and bring back as many inscribed clay tablets as possible. These he stored in a great library. His scholars constructed a dictionary and, in addition, transcribed the contents of many of these tablets into the three languages of Assyria, Sumer, and Babylon.

What do these clay tablets tell us? People who had lived as long ago as 3000 B.C. had already acquired great social and cultural capabilities. They developed a script, and by this means recorded and preserved for future generations a history of their thoughts, activities, and beliefs.

Your author is greatly indebted to a close personal friend, Dr. Menahem Mansoor, Professor Emeritus of Hebrew and Semitic Languages at the University of Wisconsin-Madison. Dr. Mansoor, through personal conversation and by supplying synopses of his lecture materials, has been of indispensable help in understanding these first, most remarkable historians.

Dr. Mansoor indicates that as far back as the tablets go, the theologians of Babylon and of Sumer never questioned the existence of a pantheon of gods and goddesses. They believed that gods were living beings with human form and behavior; that they were invisible, superhuman and immortal; that they guided and controlled the world – from the sun, moon, stars, heaven, Earth, mountain, to the ditch, the spade and the garden. Everything was under the control of one or more of the gods or goddesses. They also believed that the gods and goddesses were not all the same in dignity and rank. The god who controlled a tree could surely not be equal to the sun god. Their writings further express belief in a supreme god, similar to the king of a state or nation, who ruled over an assembly of gods. These early theologians divided the universe into four main divisions and proposed that every element of nature fell under control of one or the other of these divisions.

Dr. Mansoor is of the opinion that a story of creation figures in the beliefs of most ancient civilizations. Some of these stories are grotesque and highly imaginative. Dr. Mansoor also notes that the Babylonian creation story is one of the most interesting, of great poetic content and especially important because of its relationship to the books of the Old Testament. The tablets which contain this story of creation were discovered in 1840 but had to remain undeciphered until the cuneiform code was learned. The story consists of 940 lines on seven numbered tablets.

The transcribed title means "In the Heights," just as the Old Testament is titled "In the Beginning." The story states at first there was only chaos and water, in which lived the water gods. From them sprang other gods and goddesses. One of them, Marduk, subdued all of the others, after which he became supreme among them. He then created Earth and man. Scholars think that this creation story originated about 2000 B.C.

The Sumerians believed that man was fashioned of clay and that the gods had breathed life into his nostrils. According to the Assyrian text, man was created for one purpose only, to serve the gods and to relieve them of all menial tasks. They believed that kings could obtain anything they wanted by merely saying the word. Therefore, it must have seemed extremely plausible that the highest and the mightiest of all kings the king of the gods – after having finished his planning, needed only to utter the word and creation proceeded. The Sumerians held that heaven and Earth were created in the primeval chaos, and that they were separated by the air, which at the same time produced the moon and the stars, also made of air. They believed that the sun had derived from the moon. According to their logic, all life, including humans, came from a reunion of Earth, air, and water. Contemporary scientists do not dispute that all biological forms consist of these three components.

Another story of importance inscribed on these tablets of clay is the story of the great deluge, not much different from that of Noah and the flood. The story relates how the gods and goddesses, declaring that they had not been served well, became dissatisfied with human beings. They came to the decision to annihilate humans completely. They decided to accomplish this by calling forth a torrential rain which would completely cover the Earth for a period of seven days and seven nights. Surely, they said, no humans could survive such a deluge. But En Lil, the ruler of the universe, secretly decided to spare one man named Utnapishtin. While Utnapishtin was standing by the great wall, En Lil whispered into his ear that he should construct an ark, into which he should enter and remain until the deluge subsided. Utnapishtin obeyed, and after the seventh day sent out birds to look for land. When he had thus survived the flood, En Lil granted him immortality on Paradise Isle in the middle of the river.

There is an interesting note to the unearthing of these inscriptions. Roland Smith of London, who in his short lifetime accomplished so much in discovery and decipherment of many of the clay tablets, found a fragmented tablet containing only one half of the story of the deluge. Later, with one chance in a million, a miracle occurred and he found the remaining half.

We return to about 2700 B.C. The many stories, tales, and legends about Gilgamesh, a "contemporary" of Utnapishkin, have been inscribed on eighty-six tablets. Twenty-six were found originally, and sixty were found later. One story relates that the father of Gilgamesh was the god De Muzzi, while his mother was the goddess Ihana.

Another story tells how King Gilgamesh built a mighty wall at Uruk. It was eight miles in length, twenty feet high, fifteen feet thick, and surrounded the city of Uruk. Upon its top he constructed over nine hundred turreted towers. The wall which surrounded the entire city of Uruk had two gates, one opening towards Eridu in what is now Iraq, and the other towards Nippur. Within the city there were palaces, homes, and a religious temple known as a ziggurat. It towered into the sky and on its top was a platformed temple onto which the gods and goddesses could descend to communicate with the people. It was here, the legend states, that De Muzzi and Ihana came, once a year, in order to populate Earth with their progeny.

Another story told how Gilgamesh pleaded with Utnapishtin to give him the secret of immortality. Utnapishtin revealed that the secret was in an herb which lives at the bottom of the lake. Gilgamesh tied stones to his legs, descended into the lake, recovered the herb, and brought it back to land. But while Gilgamesh was removing the stones from his legs, a serpent devoured the herb and Gilgamesh was denied immortality.

Other tablets relate that the people of Uruk considered their kings to be divine, and should they accompany them in death they would be able to share the royal priviledge of associating forever with the gods and goddesses.

Several burial sites of kings have been uncovered. Many of them were plundered and robbed of their precious treasures; however, one at Uruk remained fairly complete. Here the king was laid to rest, with his queen and the princess nearby. All were dressed in garments of royalty. Golden helmets, meticulously worked to form stylized hair arrangements, had been placed on their heads. The king was surrounded by the symbols of his office. The queen was adorned with many beautifully designed and executed pieces of gold, and on the princess were miniatures of each and every piece. At the foot of the king there was a chariot with three horses. An entourage of some sixty people had accompanied the king in death. Such discoveries as these provide evidence that these people believed in a life in the hereafter.

Papyrus was an aesthetic as well as workable surface on which Egyptians could write in hieroglyphics. The completely bordered hieroglyphic at the top is called a cartouche, and usually denotes the name of a ruler.

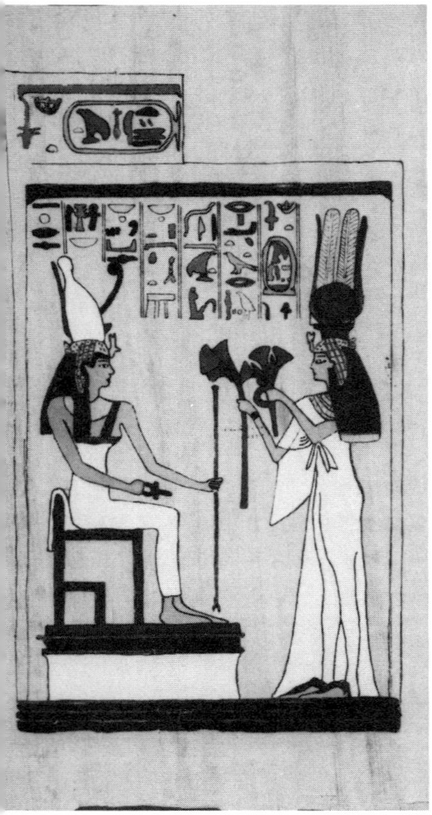

Professor Noah Kramer collected, reassembled, and studied cuneiform tablets which had been dispersed to many museums throughout the world. Had it not been for Professor Kramer, the translation of many of the stories on the clay tablets could not have been completed.

The Egyptians built their high civilization on the banks of the river Nile, for all else was desert. At first, as their earliest records tell us, Egypt was divided into the valley peoples of the south (Upper Egypt) and the delta peoples of the north (Lower Egypt). You will recall that "upper" and "lower" refer to the Nile River, which has its origin in the south and its mouth in the delta of North Egypt. About 3000 B.C., King Menes of the south conquered the delta people and united Egypt into one nation.

The early Egyptians also developed a script – a combination of pictograms, ideograms, and photograms. Records of this hieroglyphic writing remain, chiseled on stone and painted on plaster. The Egyptians developed a kind of paper made by pulping together the reeds of the papyrus plants. Documents written on papyrus are known as papyri. The Evers and Smith papyri record the medical practices of the day under the reign of the Pharaoh Imhotep. There are also early records of a thriving agriculture with elaborate systems of canals for drainage and irrigation.

From the earliest of times, dynasties of kings and pharaohs built their palaces, tombs, and places of worship along the Nile. Like the Sumerians, these people attributed divinity to their rulers, and they believed in a pantheon of gods and goddesses. But about 1400 B.C., Akhenaten, reigning pharaoh of the 18th dynasty, put an end to the practice of polytheism, and declared that there was only one god, the sun. This is the first record of monotheism. Akhenaten constructed a great temple which contained a huge gold disk, to which people bowed and offered sacrifices.

A generation later, the pharaoh Harmhab destroyed this temple. There remains, however, a stone wall leading to the temple on which was carved, in low relief, a ceremonial procession headed toward this gold disk; Akhenaten is followed by his queen, with a procession of individuals carrying gifts and sacrifices. Later, Osiris became the preserver of life, and Amut, an anthropomorphic god with the head of a lion, the body of a crocodile, and the rear of a hippopotamus, became the destroyer of the dead. Amut was thus depicted on one of the treasures, a sort of chaise, which was recovered from the antechamber of King Tutankhamen's tomb.

The many precious items which were placed in the tombs of the pharaohs, either before or after death, provide strong indication that these people, too, believed in a life after death. The persistent efforts of Howard Carter led to the discovery of the Tomb of Tutankhamen. Many of us have been privileged to view these magnificent treasures in some of our great museums. These precious items were expressions of the highest art. The traditional splendor which surrounded royalty persisted, even to the time of Cleopatra. When Mark Antony was at Alexandria, Cleopatra came to see him. What a sight it must have been to see a purple-sailed, gilded barge being propelled by Nubian slaves with silver oars, while Cleopatra, resembling Venus, reclined under a golden cloth.

Mummification, a unique method of preparing their dead, was practiced by the Egyptians. The vital organs were removed from their cavities, wrapped in linen bindings, covered with resin or pitch, and either returned to the cavities (though not always in the order in which they had been taken out, nor to the same place) or placed in a separate jar. The body was wound about by bandage-like linen strips, and these were sealed layer by layer. After this procedure the body was placed in a wooden enclosure, together with herbs and scents of various types. A stylized portrait was painted upon the wooden enclosure. One can only conjecture what the purpose was of such body preservation.

Pyramids, the sphinx, palaces and temples, lavish tombs with precious treasures, all indicate that Pharaoh was central to the lives of these people. He, his queen, and his progeny were believed to be divine and would live on after death with the gods or the god whom they had worshipped. At the same time, it seems paradoxical that the precious relics which have come down to us, which we treasure and admire today, resulted from the enslavement of people who elevated their rulers to the highest pinnacle. There were human inequalities then, as there are now.

Before proceeding, we must briefly remark about a few other civilizations. There was India, with polytheism and religious indication of life in the hereafter and of reincarnation in living form.

158 The Chinese developed a very intricate form of writing on paper made from rice stalks to record their history and legends. They also had dynasties, gods, and various rituals, and a desire to assume the perfection of their divinities.

The peoples of the New World, such as the Toltecs, Aztecs, Zapotecs, Olmecs, Maya, and Inca all had paper and alphabets and recorded histories in the form of codices. They also chiseled in stone. Unfortunately, due to the destructive efforts and burnings of the Conquistadors, much of the history and practices of these people can never be learned. However, we know that they built enormous ceremonial centers and great palaces. They performed human sacrifices to the sun god, Quetzalcoatl. Their rulers also were regarded as divine.

They, too, had their story of creation. It is there for all the world to see, carved on a huge stone wall at one end of the ceremonial center at Teotihuacan in Mexico. Tlaloc is weeping tears, which fall to the ground and there create the ocean, in which snails, shellfish, and reeds are growing. From this ocean man and the sun god, in the form of the plumed serpent, emerge.

The New World peoples were artistically advanced, as their many beautifully executed stone sculptures and the intricacies of their gold and metal workings show. Unfortunately, much of the latter was captured by the Spanish conquerors and melted into ingots for transport back to Spain. The New World people developed an intimate knowledge of astronomy; the exactitude of their calculations is recorded on calendar stones. They, too, believed in the hereafter, for they buried their dead in tombs accompanied by precious personal items and by food. Their tombs can be seen at Monte Alban in Oaxaca, and at Palenque, at the base of the Temple of the Inscriptions. Their elaborate burial preparations were similar to those of the Mesopotamians and the Egyptians.

These are only a few examples of how recent civilizations which were geographically widely separated from each other developed along somewhat similar lines, their theologies, philosophies, arts, recorded histories, stories of creation, notions of divine royalty, and beliefs in the hereafter.

We shall now look at yet another civilization which has had a profound effect upon our western culture. In ancient times, there were people who dwelt in Palestine and who were called Yehudi, meaning "man of Judah." At first this designation was given to those who belonged to the tribe of Judah, but later on it applied to all of those who lived in the kingdom of Judah. After the fall of Babylon, all of those who accepted the Jewish faith were included.

Abraham of Chaldea, in the land of Ur, was the first patriarch, and the progenitor of his tribe. He was known as a Hebrew, not as a Jew. He was succeeded by his son Isaac and in turn by Jacob, who was known also as Israel, and whose twelve sons founded the twelve tribes of Israel. One of Jacob's sons, Joseph, was captured and taken as a prisoner to Egypt, where he became a confidant of the Pharaoh and rose to a position of eminence. Jacob and his other sons went to Egypt and settled in the land of Goshen. Here they were reduced to slavery and were finally led out by Moses into the promised land, where under Joshua they conquered Jerusalem.

In the 11th century B.C., democratic Israel became a monarchy under King Saul. In 977 B.C., after the death of King Solomon, the people split up into the ten tribes of Israel and the two tribes of Judah. The tribes of Israel were later captured by Shalmanezar and taken to Assyria, after which they became lost to history. In the year 586 B.C., Jerusalem was destroyed and the people of Judah who survived were taken to Babylonia by Nebuchadnezzar. Then in 530 B.C., King Cyrus of Persia captured Babylon, but he permitted the people of Judah who wished to do so to return to Jerusalem. There they lived until 70 A.D., when the city was again destroyed, this time by Rome, and the people of Judah were taken into slavery.

The Jews who had remained in Babylonia set up four schools. In 948 A.D. these schools were closed and those Jews who lived there were dispersed to many European countries. During World War II, under Hitler, many of them suffered the greatest of all human tragedies, that of mass extermination.

At last in the year 1948, the United Nations decreed and ratified that the state of Israel should be created as a national identity and homeland for the Jewish people. The Jews have produced great historians, theologians, and scholars, and they have made wonderful contributions to literature, music, science, and the visual arts. They have built magnificent temples to the one god whom they serve. They had, apparently, worshipped him with dignity from at least as

A page from the Dead Sea Scrolls.

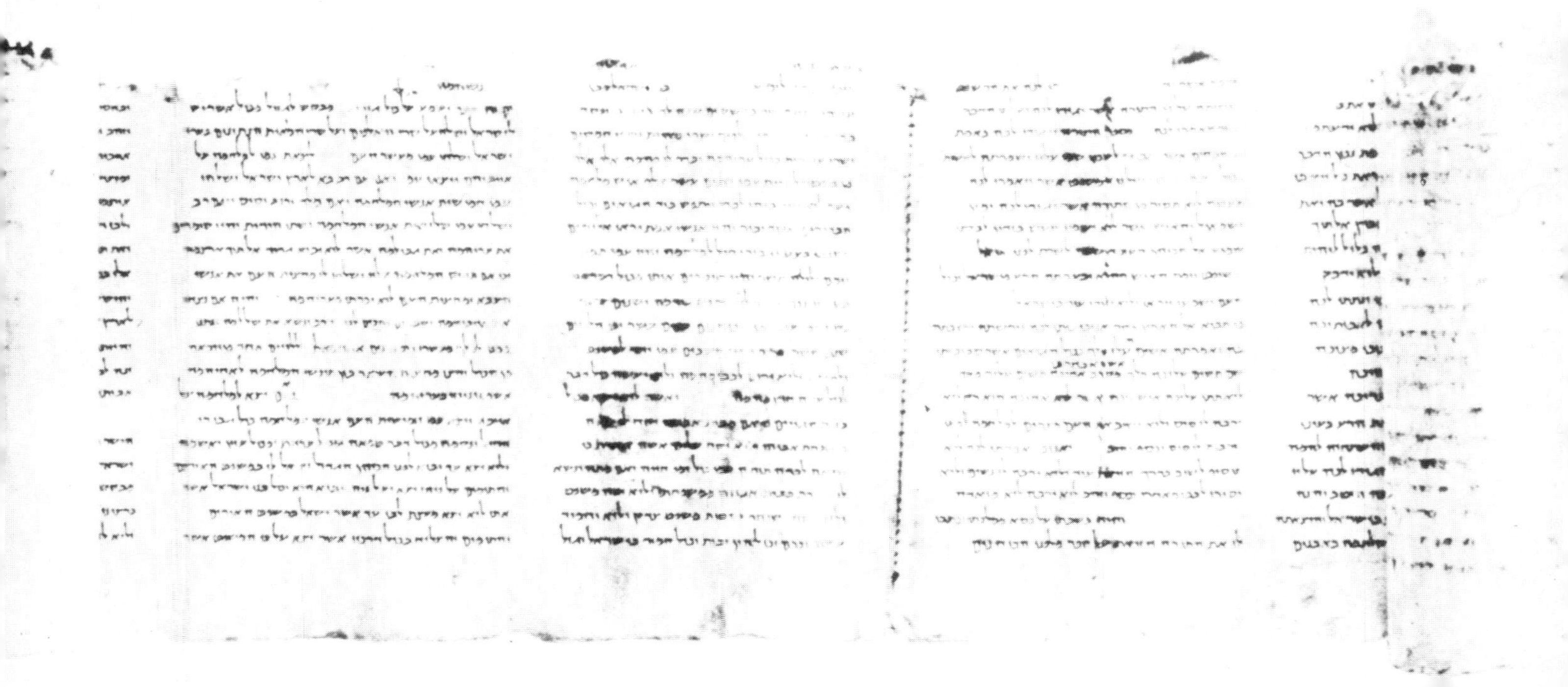

far back as 2000 years B.C. This is indicated by the previously described intricately designed and executed ceremonial crowns, maces, and standards which were found in a cave overlooking the Dead Sea.

The discovery of the Dead Sea Scrolls has further added to the recorded history of the Jewish people. One such scroll consists of seven pieces of leather which were sewn together, on which there are writings in both Aramaic and Hebrew, including the entire book of Isaiah. Carbon-14 tests have determined that this scroll was inscribed at some time between 200 years B.C.and 100 years A.D. Many other scrolls contain records of Jewish and early Christian history and have helped to establish a continuity from ancient Mesopotamia throughout Jewish history into our modern theological and philosophical concepts. Indeed we owe a great deal to the heritage of the Jewish people, who derived much from what had been inscribed upon the tablets of clay from the land of Ur, where they had their origins.

It is hoped that these few brief historical references will leave the impression that there were people who had hopes and aspirations, a belief that they had been divinely created, and that after death they, themselves, might become immortal. People of widely separated civilizations simultaneously arrived at such concepts. But what might our mythical travelers think about all of this? How have these observations affected their reasoning and logic? To what conclusions might they have come? They have about 2000 years to go before their journey comes to an end. They will have many more encounters which may temper their thoughts.

Upper) - A cluster of caves on the northwestern shores of the Dead Sea where several thousand scroll fragments were uncovered. Lower) - The Shrine of the Book at the Israel Museum in Jersalem houses the Dead Sea Scrolls, among them the Isaiah Scroll.

Part 4: HUMANS COME TO PLANET EARTH

Chapter 29: MYTHOLOGIES, RELIGIONS, SCIENCE, AND HUMAN CONCEPTS

A hermit sits before the entrance to his shelter and views the world from a very limited perspective. Others build symbolic high walls and cloister themselves within, so that their knowledge can never extend beyond. But there are still others who pursue every avenue, pathway, and dimension of the boundless universe to better understand who and what they are and why and how they have evolved. This is what our mythical travellers are doing. In their journey from the very beginning of cosmic time, their views are limited only by the capabilities with which humans are endowed.

They are about to continue their observations into the last two and a half millennia of Earth's existence. They will meet some of the minds who formulated ideas, beliefs, and actions far more comprehensive than those of any previous time.

About the beginning of this epoch, something akin to intellectual restlessness began to make its appearance. The scientific approach as a means of obtaining answers to many perplexing questions was awakening. Mythological explanations for the creation of the universe and its workings were coming to be recognized as somewhat less than adequate; however, it is difficult to know just when departure from belief in the myths and legends of old occurred.

Since 3000 years B.C., the Mesopotamians demonstrated their abilities in advanced mathematics. Clay tablets, now at Yale University, depict equations in cuneiform symbols. Drawings on the tablets present the formula for arriving at the square root of the number two. This was discussed in the June 1980 issue of Scientific American and again in two letters in the April 1981 issue. These tablets date from many centuries before the same problems were tackled by the great Greek mathematicians, Pythagoras and Euclid. The Egyptians, also, were adept at mathematics and medicine, as recorded in the Smith and Ebers papyri.

In the third century B.C., the Greek Eratosthenes, traveling in Egypt, made an astounding discovery. During his stay in Syene, now Aswan, he learned that at high noon on a given day, when a stick was placed vertical to the surface of the Earth, the sun cast no shadow. However, in the city of Alexandria, 500 miles to the north, on the same day and at the same time, the sun cast a definite shadow from a vertical stick. By measuring the length of the shadow, and knowing the distance between these two cities, Eratosthenes calculated that the circumference of the Earth was 50 times 500 miles, or 25,000 miles. This is fairly close to the sophisticated measurements of the present day.

The golden age of Greece offers some of the best examples of thought development. Prior to this time it was thought that humans were created by the gods, were inferior to them, and were made only to live in servitude and to perform the menial tasks which were beneath godly dignity. A variation of this was that Prometheus, the Greek demigod or hero, fashioned man from water and from earth, much as in the biblical story where man is made from dust

Another legend indicated that Cadmus had been instructed by the virgin goddess Athena to scatter serpents' teeth over the surface of the Earth, wherefrom a force of armed men had arisen. Other gods and goddesses while walking on planet Earth had cast stones over their shoulders and from these humans developed. Some sources attribute this legend to Deucalion and Pyrrha, humans who were survivors of the great flood.

The legends of the Abron, native to New Guinea, tell of a sacred cave in which the gods fashioned, in stone, a white man and a white woman. From these all of the Abron descended.

Many other myths and legends contain stories of creation. A fanciful and improbable point of view was expressed by the Greek Empedocles, who lived between 495 and 435 B.C. Empedocles wrote that certain parts of humans and of animals (such as eyes, heads without necks, arms without shoulders, etc.) had been floating about in space and that individual parts or groupings had joined together, making humans and animals as we know them today. Other parts had grown together in an irregular fashion, producing the occasional monstrosity. Now and then an animal part would join with a human part to produce the mythical creatures such as the centaurs, the Minotaur, and the sirens or mermaids. He stated that these later became extinct because they could not reproduce themselves.

Shortly before the time of Empedocles, another Greek, Anaximander, who lived between 611 and 547 B.C., came upon the scene. He had been a pupil of Thales, who is credited with having been the first materialistic philosopher. Anaximander reasoned that the universe and its components had arisen from an indefinite, undifferentiated, ever present substance which had somehow sifted out itself to form the Earth, the sky, the sun, the moon, and the stars and, in addition, had also formed hot and cold, moist and dry. He called this matter, which was in constant motion, "the infinite." This early thinker believed and taught that life originated in water, originally resembled fish or had been born within fish, and had come to reside on land only when the creatures found that they could obtain nourishment there. From these philosophies, later scientists found a few thoughts on which to build.

What did some of the early sages think about social and cultural advances? Travel began to influence thought. It became apparent that people who lived far from each other developed somewhat different social and cultural patterns. One of these travelers, Herodotus, who lived between 484 and 425 B.C., visited over 50 different peoples. He observed and recorded in detail their food, clothing, actions, marriage customs, and beliefs. He then made comparative evaluations of his data and came to the conclusion that each civilization had arrived at the cultural and social status that appeared best for its individual purposes. He was of the opinion that his gods and those of others possessed the same basic functions, even though they had been given different names.

Democritus, a pupil of Herodotus, lived between 460 and 370 B.C. His insight into the nature of things is indeed almost incredible. He wrote that all reality consisted of atoms – similar, indivisible particles differing only in shape, size, and velocity. The universe and all that it contained came into being by the impact of such atoms, Democritus said, and their combinations produced an infinite variety of forms. Further, he stated that man himself and even his mind had a material basis. Democritus went even beyond this. He declared that since humans were endowed with reason, they could choose those things which were beneficial to them. These choices were of their own making, rather than having divine origin. He taught that humans developed certain of their social and cultural customs, such as group living, the tribe, civilization, language, education, and the arts, because of their own needs and by their own wills.

The Skeptics doubted ideas which had been previously accepted, especially those having to do with religion, divinity, and the views of creation related in the myths and legends. Protagoras, one early Skeptic, lived during the golden years of Greece between 490 and 421 B.C. He declared in *Truth and Rejections*, "Man is the measure of all things," and "I have no means of knowing whether gods exist or not." He was perhaps the first agnostic, for he could not fully accept nor totally reject ideas of divinity, because he had no proof that their legendary influences were true or untrue. He was of the opinion that social and cultural patterns had evolved from an original homogeneity to the complex status of his time. He conceded, in his writings, that the gods had been of some benefit to man and should be worshipped whether or not one actually believed in them. Others of his time considered religion to be useless and worthy only of being totally discarded.

Gorgias, for example, a contemporary of Protagoras, wrote of nihilism,

 "Nothing exists; if anything did exist, it could not be known; and if it were known, it could not be communicated." Still others rejected these ideas. Socrates, who lived between 470 and 399 B.C., taught that there were universal values and principles in all societies which ultimately regulated the evolution of social and moral justice. But such values and principles could not be readily perceived by the human mind, because people persisted in remaining ignorant of them. Plato wrote in *The Republic* that all societies eventually specialized, dividing themselves into three groups: the producers, the protectors, and the rulers. He stated that humans are endowed with certain innate values which are the basic reflections of universal laws. These were the values taught by Socrates. Plato declared that they had been given to man by his divine creator, because man was a part of nature and of the universal whole.

Between the years 384 and 322 B.C. there lived a most remarkable man named Aristotle. During his lifetime, particularly because of his insatiable curiosity and quest for knowledge in practically every field of science and philosophy, learning advanced far beyond that of any previous period. He was tutor to Alexander the Great, the Macedonian King, through whose influence he was able to establish his school, the Lyceum, in Athens. His teachings and writings encouraged the youth of Athens to drift away from the teachings of the great theologians. Soon after the death of Alexander the Great, the Lyceum was closed. Artistotle retired to his native village never to teach again, and died a year later. Even though some of his conclusions were later found to be in error, the impact of this one individual on all future scholarship and learning was tremendous.

The Stoics did not accept that there had ever been divine intervention in human affairs. One of them, Epicurus, taught that all advances within a society or culture were brought about by the necessary requirement for protection and survival, and that the search for a good life constituted the only desirable end.

With the decline of Greek influence, the scene shifted to Rome. Lucretius, who lived from 99 B.C. until 55 B.C., foretold many of the basic principles of the scientific future. The impact of his wisdom is indicated by this quote from his poem "The Nature of Things."

> The universe was created by no tool of god but by a confluence of atoms, the primal germs of things. The first plants and animals arose by spontaneous generation from the warm moist soil. Man and his soul both had material origins. At first, men lived like beasts without clothing or tools, gathering berries and acorns as food. Later, arming themselves with clubs they hunted animals, eating the flesh and using the skins to cover themselves. They lived in sexual promiscuity. But the invention of hut building made men and women cohabit, thus marking the beginning of family life.

The materialistic ideas of Lucretius, who believed that humans were responsible for their own social orders and their own destinies independent of any divine intervention, prevailed in Rome until the death of Marcus Aurelius in 180 A.D. During those years Rome was beset by invasions, wars, and plagues. People were deprived and civil war resulted. Their minds were ripe for change, for hope of security and happiness, not necessarily in this life, but perhaps only in the hereafter.

Consequently, many cults arose which promised them what they sought. Banned by the authorities, they nevertheless seemed to flourish. Among the most prominent of these cults were those of Orpheus, of Osiris and Isis, of Mithra, and of the mother goddess Cybele, in which a kind of baptismal rite was performed. In all of them the observance of certain ritualistic procedures and the holding of certain beliefs would guarantee believers a glorious eternal life in the hereafter.

It was about this time that Jesus Christ of Nazareth appeared. During his 33 years of life, he proclaimed himself to be the Son of God and at one with Him and the Holy Spirit. He preached the universal brotherhood of man. He set about establishing principles of morals far beyond anything which had been previously proclaimed. He was believed to be the Messiah who had been historically prophesied. He was the Savior who would be crucified to redeem the sins of man. To those who believed, he promised that there were mansions in the sky where after death they would be far removed from the sufferings of this world. The unusual circumstances of his virgin birth and of his physical ascension into heaven were cited as proof that indeed he was supernatural, that he was God.

The cult of Christianity was also banned as illegal. However, despite the threats of persecution and martyrdom, there were those who believed so firmly in Christ that they traveled widely and taught all that they knew about him. St. Paul preached before huge crowds in Antioch, Tarsus, Damascus, Cyprus, Athens, Corinth, and in many other places. He was often joined by St. Barnabas. Multitudes began to accept these beliefs as their own. Slowly but surely Christianity was on its way to becoming one of the major religions of the world.

Even though some historians consider Christian teachings to be a continuum from the philosophies of the earliest Greek humanists, the high moral code and the promises of eternal happiness in a life hereafter were perhaps the factors which caused Christianity to flourish.

It was not until the 4th century A.D., during the reign of Constantine, that Rome declared Christianity to be its official religion. Unfortunately, as time passed, many of those who reached high positions as theologians learned that their prestige and power could be used as tools in the political realm, and at times rulers and theologians became entwined in the most diabolical schemes.

St. Augustine (354-430 A.D.) established the principles of Christian dogma which persisted for many years to come. In his book *The City Of God* he declared that God was the omnipotent and perfect creator of the universe and all that it contained, including sinful man. Humans could never fully understand God's creativeness and purpose. They were damned to sinfulness, depravity, and suffering from the time that Adam and Eve departed from the will of God in the Garden of Eden. Humans were ignorant and incapable of making inquiry, scientific or otherwise. Augustine taught that the scriptures revealed everything to humans which God intended them to know. Despite his theocentric orientation, St. Augustine had a great appreciation for the value of history, in that important people and unique epics had constituted the creative drama of human existence. He influenced his pupil, Horotius, to write a comprehensive history of the barbarians and of the pagans who had invaded Rome. During the next several centuries, many individuals, including Isador of Seville, followed this example and wrote extensively of other nations and peoples.

St. Thomas Aquinas, who lived between 1225 and 1274 A.D., held ideas which were somewhat different from those of St. Augustine. He rediscovered the works of Aristotle, and his own ideas became deeply influenced by them. He declared that humans are intelligent, that they are endowed with reason, that these gifts were granted to them by their divine creator and that they are duty-bound to use them for the acquisition of knowledge regarding things both natural and theological. This view differed radically from previous Christian teachings, and a few years following the death of St. Thomas Aquinas the works of Aristotle were banned by the high theologians. Nevertheless, under the influence of Aquinas, the stage seemed to have been set for the application of reason and logic to both scientific and theological matters. Roger Bacon, a professor at Oxford and contemporary of Aquinas, also taught that reason and experience were the only means by which truth could be ascertained.

The theological philosophy of divine intervention in human affairs was not confined to the God of Abraham, nor to the three-in-one God of the Christian discipline. In the 7th century A.D., Mohammed proclaimed that there was only one god, named Allah, and that he, Mohammed, was Allah's prophet. The Islamic beliefs are now held by most of the peoples of the Near East and Asia Minor, and in large parts of Africa and Asia.

Hinduism arose in India, and Buddhism gained prominence in the Orient. Great nations and groups of nations have declared their beliefs to be superior. Regardless of how high were their moral concepts, great wars have been fought to preserve the individualities of these beliefs.

In the physical sciences, the age of the universe and of planet Earth are now well documented, as are the chemistry and the physics involved in their creation and their ever-changing evolution. In the biological sciences, there has come about an awesome nearness to the basis of life itself. In the realm of mathematics, computerization has permitted humans to travel to the moon, to land, and to return again to Earth, and has made it possible for unmanned vehicles to explore the outermost planets of our solar system. Scientific achievements have made great advances in agriculture, in plant and animal genetics, in all forms of industry, in more and more rapid forms of transportation, and in the healing arts. All of these have come about by human ingenuity.

How should the human mind approach seeming conflicts between mythology, religion, and science? Religion cannot be delved into by scientific methods. Science is a process which can be subjected to positives or negatives; theology is a philosophy of the mysterious which can be accepted only on the basis of belief and faith. As far back as written history goes and up to the present time, most humans have had such beliefs and faith in a divine planner, creator, and overseer of the universe and all that is within it. Most scientists are of the opinion that while they know much about the universe and its chemistry and physics, and about the electromagnetic and gravitational forces which enable it to perform all of its functions in the most magnificent, mathematically precise manner, that which is unknown far exceeds that which is known. There are certain aspects of the creation and evolution of the universe which are not within the range of human capabilities ever to resolve.

Part of the conflict, then, between religion and science seems to be between those who believe that the Bible is the revealed word of God from which no departure can be made, and those who hold that the Bible was written by men and in symbolic terms. Fundamentalists who cannot stray from Biblical teachings believe that we have all of the answers; they reject the efforts of those working in the scientific realm as not only useless but anti-theological as well. It is not the purpose of these writings to dictate to our mythical travelers, nor to our readers, those things in which they should believe. All of us are endowed with the capacity to reason and to make choices. In light of this capacity, we have the responsibility to reach our ultimate conclusions on the basis of enlightenment rather than ignorance, and in conformity with our own individual concepts of truth.

As these travels end, let us reflect on all that we have observed and experienced, and conclude, as best we can, with our own beliefs.

Early Greek Philosophers

Protagoras	490-421 B.C.	Taught that all perceptions are true. He believed that opposite qualities are copresent in objects, and seeming conflicts arise because different people perceive different qualities. Yet some perceptions are better than others; therefore, teachers must make what is just or better seem true to their students.
Socrates	470-399 B.C.	Was the turning point in Greek philosophy, and the orientation for most subsequent philosophers. He is most famous for the "dialectic" method of questioning what others believed they knew as the truth. He thereby proved that neither he nor they knew the truth; but by questioning, both were inspired to search for it. He was condemned by the government for advocating criticism of all institutions, including the government. Yet, he was neither an anarchist nor an atheist.
Democritus	460-370 B.C.	One of the earliest Greek Atomists, he believed that atoms and the void are alone real. He taught that objects of sensation exist only by convention, and that perception is not trustworthy. He advocated critical reflection on the evidence of our senses as the best approach to truth; yet believed that even legitimate knowledge is, at best, only opinion.
Plato	428-347 B.C.	A pupil of Socrates, he put all of what we know about Socrates into writing; in fact, he wrote extensively on a wide range of subjects. He is famous for the theory of forms--that all we know is what we assess against an ideal. He taught that all plurals are measured against that which is singular, ideal, and defined.
Aristotle	384-322 B.C.	Wrote extensively on popular, philosophical, and scientific subjects. Virtually all of his writings concern the quest for the answer to the question, "what is substance?"
Epicurus	341-270 B.C.	Believed that feelings of pleasure and pain that accompany sensual experiences are the ultimate good and evil, and that all knowledge derives from sensory experience. He taught that the body and soul are separate, but cannot exist without each other. He is known for advocating a way of life that involved cultivating friendship, enjoying harmless pleasures, and tranquility.
Lucretius	99-55 B.C.	Believed that mythology was artifice, and that state religion served only a political purpose. He advocated nature worship--the belief that one can reject all trappings of religion and explain all natural phenomena on a purely materialistic basis, and yet retain reverence and adoration. He also believed in "natural justice," that is, that in the past social groups had made bargains for mutual profit not to hurt or be hurt. Any social groups that didn't make such a pact didn't survive.

EPILOGUE

Strange is it not that of the Myriads
who before us passed the door of darkness through,
not one returns to tell us of the road
which to discover we must travel too.

Rubaiyat of Omar Khayyam

Dear Readers:

The unfolding of endlessly wondrous events on the stage of cosmic time and space has brought before us the drama of life in its entirety. It has been an intense drama, because it has dealt with pieces of everything that ever was, is now, or ever shall be. The many scenes of this saga have attempted to portray for our mythical travelers, who are really you, our readers, the many encounters, the many progressive sequences, which have led not only to the creation of the universe but to the coming into being of planet Earth. We have followed Earth's preparation for the advent of certain conglomerates of matter possessed of that mysterious quality called life.

Other parts of the drama have shown how Earth, aided by energy waves of heat and light from the sun, became not only able to sustain the life which had been created but caused it to grow, function, reproduce, adapt, and even modify its structure in order that it might continue to function in an everchanging environment. Biological forms which could not meet this requirement perished, while those which could adapt and modify moved forever forward.

During long periods of evolutionary time, the various scenes of this play depicted greater and greater complexity of structure, more and more refinements of function, until at last, about 35,000 years ago, creatures came upon the scene who were indistinguishable from human beings who live now, in the 20th century.

This seemingly was the attainment of the ultimate goal of creation, for *Homo sapiens sapiens*, the wisest of the wise, became dominant over all other living creatures. Humans were endowed with certain advanced characteristics. They could perceive, and they could integrate their perceptions. They could memorize them and recall them when it suited their fancies. They could reason, they could make choices, and they could even make inquiry into the why and wherefore of their being.

They used all these characteristics to create, from the beginning of their time, certain social and cultural patterns. From primitive stone tools, which in all probability had their first manufacture and use quite by accident, humans have progressed to a high order of intellectuality in the realms of science and technology. They have probed into the very nature of the universe and of its components. Humans have indeed made great progress in acquiring knowledge of the chemistry and the physics of life itself.

Humans have been both inventive and creative to meet the demands of their own material needs. Humans of the 20th century can be transported rapidly over land, on and under the waters, and through the air, so that remote destinations on Earth are only hours away. With manned lunar landings and unmanned exploration to the outer planets of our solar system, travel throughout the universe is beginning to come under human control. Indeed, there are even some serious thoughts about establishing colonies in outer space and possibly on other planets, should population growth on Earth so require.

Humans have composed marvelous music and devised instruments and combinations of instruments so that such music can be enjoyed by all. There have been great sculptors and painters and architects who have conveyed their own particular messages in their individual ways. History has been recorded, and the great classics of literature and poetry have been preserved. All of these are readily available in numerous museums and libraries.

In many nations, great universities have arisen. In them, research, teaching, and learning have been dedicated not only to the acquisition and expansion of knowledge but to its application, attempting to resolve the many problems which seem to be a part of human existence. Science and the humanities go hand in hand in these endeavors.

These are things which are within the human potential to know. What about those things which are beyond human comprehension? What about life itself, which cannot be analyzed nor expressed in definitive terms of substance and matter? And this indefinable force called gravity, which in some mysterious manner keeps the various components of the universe in their respective orbits and spatial relationships? And the mathematical exactitude of the electric potential of each and every atom? And what about the wavelengths of heat and light, where particular matter becomes invisible and is expressed in terms of energy? And this same light, which makes it possible for us to view the beauty of color and form which surrounds us, and without which photosynthesis could not create the food chain which sustains us? And what about the intellect, the mind, the soul, those centers of the highest potential of human expression?

What are the relationships between these things and the material substances in which they are housed? The detailed understanding of all of these is beyond human comprehension, and yet it is known that they are manifest everywhere and that they do exist. They occupy material substance and yet they are not material. They belong to separate categories which are beyond human definition. One example: when a human being is deprived of that mysterious quality called life, the material substance is still there, for a time at least, but the functions have ceased, the brain can no longer provide the mechanism by means of which the intellect, the mind, and the soul operate. Since these characteristics, which are essential to the lower, intermediate, and higher processes in humans, are beyond human understanding and human comprehension, is there a possibility that their exists a "superhuman" greater than ourselves, who does comprehend these human imponderables?

The majority of humans throughout the world firmly believe that there is an invisible, all-powerful, omnipotent, supreme, and divine being who not only designed but also created and continues to oversee the universe and all that it contains. They also believe that, in some manner, these expressions of life which occupy the material body but, which in and of themselves are not material, will continue to exist in a life hereafter. As far back as historical records go, which is about 5,000 years, humans have had such convictions. Many religious writings state that such divine beings or gods appeared and become visible to humans. Utnapishkin in ancient Mesopotamia had such an experience. The story of Abraham and his covenant with God is well known. And in the Christian discipline, God became man.

Science and religion cannot be probed by the same tools. In science, when there are all positives and no negatives, hypothesis becomes factual, a scientific truth. However, within the theological realm such approaches are totally inadequate. If humans are to believe that a being greater than themselves exists, then they must do so on the basis of faith, and of faith alone. To paraphrase Omar Khayyam in The Rubaiyat, of all those who have passed before, not one has ever returned to tell us about the mysteries which humans have failed to penetrate.

We previosly mentioned that St. Augustine taught that humans were sinful and ignorant and that God had revealed in the Biblical writings all that He ever intended humans to know. All of the answers were contained therein, and it was therefore useless for humans even to try to seek answers elsewhere.

They must accept the absolute truth of these Biblical writings and depart from them not at all.

Today in this, our 20th century, many people abide by the teachings of St. Augustine. They consider the scientific approach to be anathema and believe that nothing need be known other than that contained in Biblical writings. However, in the 13th century A.D., St. Thomas Aquinas taught and wrote that since humans were endowed by their God with reason, logic, will, and choice, they were duty-bound to use these faculties to search for the truth.

Many have followed the teachings of Aquinas, yet today there are numerous theologians who continue to teach the admonitions of St. Augustine. Others have gone far in trying to resolve the relationships between science and theology. There are still others, intermediates in both the realms of science and religion, who accept some things but deny others.

So the question remains: how can human aspirations and beliefs in a divine being, which have persisted throughout all known history, be evaluated? The magnificent universe, the abundant Earth, the mind, and the soul, together with all of the knowledge which has been acquired through the scientific approaches, must be taken into account in arriving at any final conviction.

Have humans attained perfection? The answer can only be that indeed they have not. Despite their many great achievements for good, they have also been guilty of much that is evil. Hatred is rampant over love. Humans have failed to tolerate and to understand the beliefs and actions of others which differ somewhat from their own. Color and race are factors in the mistreatment of others. Great industries pollute our land, our waters, and our atmosphere. Humans vandalize, waste, burn, and destroy precious irreplaceable resources. They persist in aggression and engage in hideous crimes of arson, rape, and murder. Humans originate great wars, and kill each other by the millions. They have not learned that such actions have never solved the problems which they were designed to settle. Finally, humans have devised the most diabolical of all schemes, that of the nuclear weapon, which if used in sufficient numbers could destroy all living things, and could return Earth to the sterile thing it once was. Indeed, humans are far from perfect.

But will there ever evolve during some epoch that lies ahead a *Homo sapiens futurensis* who will learn to avoid these evils and to work for the benefit rather than the detriment of humans? Will there ever evolve a creature who will teach the young that their immense heritage deserves something better than endless and continuous trivia, nonsense, crime, and pornography with which our three important education and communicative media – the press, radio, and television – confront them? Will future humans be able to convince their leadership that wars and killings are useless in trying to solve the great problems of the world? And will future generations abide by the great moral principles which could guide humans toward true perfection?

Let us not abandon hope, for within the human mind and the human soul there resides the potential to go forth in overcoming human defects. There can be, if humans will it, an ever-expanding trend toward perfection of their kind. Let us hope that each generation of Begetters will produce Begotten more nearly perfect than they. Perhaps through a long period of evolution, generation after generation will cause the high ideals of science and theology to fuse, permitting the ultimate reality to emerge.

The play has ended. The final curtain has fallen upon the many scenes which have portrayed the past and projected the future. Our mythical travelers and you, our readers, are left with your thoughts on all that you have observed and experienced. What now have you to say about this? What are your convictions? What do you believe, and what beliefs or ideas are you willing to discard?

It is not the purpose of these writings to make these decisions for you. You and all other humans are individuals, each and every one. Your conscience, your mind, your soul are your very own, and your decision should be based on enlightenment rather than ignorance, in conformity with the truth as you find it to be. If this play, staged as it has been from the beginning of cosmic time until this present 20th century, has aided you in any way in arriving at some definite conclusions, then the message will have been conveyed and the purpose fulfilled.

Insofar as possible, your author has sincerely tried to suppress any personal bias, in order that you not be influenced by his point of view. It is hoped that our travelers, and you, our readers, will have profited from these writings as much as has your author. It has been a rewarding undertaking to set forth the results of my own experiences and thinking, to collect materials, to analyze them, and to build around them various philosophies. My deep gratitude goes to all who peruse these words and ponder these thoughts.

Sincerely,

Thomas A. Leonard, M.D.

PICTURE CREDITS

Grateful acknowledgement is made to copyright holders for permission to reproduce the following:

p. *iii*: From the collection of Dr. Thomas A. Leonard.
p. *vi*: Statue from the collection of Professor and Mrs. Michael Petrovich.
p. *xii*: Hannah Manthey appears by permission of William and Judy Manthey.
p. 1: University of Wisconsin-Madison College Library.
p. 2: Kirk Bogle.
p. 10: From Paul A. Tipler, *Physics*, Second Ed., Worth Publishers, New York, 1982
p. 11: University of Wisconsin-Madison College Library.
p. 12: University of Wisconsin-Madison College Library.
p. 15: Courtesy Department Library Services, American Museum of Natural History.
p. 25: From Colin A. Ronin, *Galileo*, Weidenfeld & Nicolson, 1974.
p. 27: From Robert A. Anderson, *Science*, vol. 148, p. 1189, 28 May 1965. Copyright 1965 by AAAS.
p. 30: Kirk Bogle.
p. 35 (upper): From L.P Wisniewski & K.Hirschhorn: "A Guide to Human Chromosome Defects," 2nd ed. White Plains, NY: The March of Dimes Birth Defects Foundation, BD: OAS 16(6): 29, 1980, with permission of the copyright holder, the March of Dimes Birth Defects Foundation; (lower): William Tai.
p. 39: University of Wisconsin-Madison Geology Museum.
p. 42 (upper): © David Scharf - Peter Arnold, Inc.; (middle & lower): Eric Grave, New York.
p. 43 (upper): From Williams & Fisher, *An Electron Micrographic Atlas*. Courtesy of Charles C. Thomas, Publisher, Springfield, Illinois; (lower): Lee Simon.
p. 44: From Peter J. Russell, *Genetics*. Copyright © 1986 by Peter J. Russell. Reprinted by permission of Scott, Foresman/Little, Brown College Division.
p. 45: Eric Grave, New York.
p. 49: From Helena Curtis, *Biology*, 4th ed., Worth Publishers, New York, 1983.
p. 53: From Peter J. Russell, *Genetics*. Copyright © 1986 by Peter J. Russell. Reprinted by permission of Scott, Foresman/Little, Brown College Division.
p. 55: Chris Hartmann.
p. 57: From Peter J. Russell, *Genetics*. Copyright © 1986 by Peter J. Russell. Reprinted by permission of Scott, Foresman/Little, Brown College Division.
p. 60: From *Journal of General Microbiology*, vol. 40, 1965. Reprinted by permission of Society for General Microbiology.
p. 62: John T. Bonner.
p. 71: From M. Auroux & P. Hagel, *Illustrated Human Embryology*, Vol. 3, Masson, 1974.
p. 74: University of Wisconsin-Madison Geology Museum.
p. 75: Carnegie Museum of Natural History.
p. 76: University of Wisconsin-Madison Geology Museum.
p. 77: From the collection of Dr. Thomas Leonard.
p. 82: Photo by Ralph Buchsbaum.
p. 83: From Robert Barnes, *Invertebrate Zoology*, 4th Ed., Sauders College Publishing.
p. 86: Edward S. Ross.

p. 88: Photo by Ralph Buchsbaum from *Living Invertebrates,* Blackwell/Boxwood Press.
p. 89: Kirk Bogle.
p. 91: Kirk Bogle.
p. 94: U.S. Department of the Interior, National Park Service, photograph by Richard Frear.
p. 95: University of Wisconsin-Madison Geology Museum.
p. 96: University of Wisconsin-Madison Geology Museum.
p. 100: From Helena Curtis, *Biology*, 4th Ed., Worth Publishers, New York, 1982.
p. 108: From Helena Curtis, *Biology*, 4th Ed., Worth Publishers, New York, 1982.
p. 116: Vatican Library.
p. 117: From Helena Curtis, *Biology*, 4th Ed., Worth Publishers, New York, 1982.
p. 126: Institute of Human Origins.
p. 132: Courtesy Department Library Services, American Museum of Natural History. Photograph by Charles H. Coles.
p. 140: Peabody Museum, Harvard University, photograph by Hillel Burger.
p. 142: Courtesy Department Library Services, American Museum of Natural History.
p. 143: From *Human Evolution* of the NSF Mosaic Reader Series, Avery Publishing Group, Inc , Garden City Park, New York.
p. 144: Courtesy Department Library Services, American Museum of Natural History.
p. 146: University of Wisconsin-Madison Geology Museum.
p. 152: From the collection of Menahem Mansoor.
p. 156: From the collection of Dr. David E. and Mrs. Marjorie Bartlett.
p. 159: From the collection of Menahem Mansoor.
p. 160: From the collection of Menahem Mansoor.